W0254444

Economic Reforms and Small Farms

Published by
ACADEMIC FOUNDATION
in association with

Institute for Human Development, New Delhi

AUTHORS

Parmod Kumar is presently Professor and Head, Agricultural Development and Rural Transformation Centre, Institute for Social and Economic Change, Bangalore. Previously, he worked at the National Council of Applied Economic Research, New Delhi. He obtained his Post-doctorate in Economics as 'Sir Ratan Tata Fellow' from the Institute of Economic Growth, Delhi and Doctorate in Economics from the Jawaharlal Nehru University, New Delhi. He also worked as a faculty at Delhi University in his initial academic career. He was Fellow under the 'International Visitors Program' sponsored by the Government of United States.

Dr. Kumar has authored several books and published more than 30 research articles in the refereed national and international journals. His core theme of work relates to agriculture and rural development. He is leading several research projects sponsored by the Government of India and various international organisations.

Sandip Sarkar is currently Professor at the Institute for Human Development, New Delhi. Previously, he has worked in several research institutes like Institute of Economic Growth and Institute for Studies in Industrial Development. He is a doctorate in Economics in the field of agro-industry and its inter-linkage with agriculture.

He has authored two books and large number of research articles in reputed journals. His main area of research is poverty, labour and livelihood in agriculture and non-agricultural sectors in which he has worked over two decades. He has been extensively involved in several large research projects funded by reputed national and international agencies.

Economic Reforms and Small Farms

Implications for Production, Marketing and Employment

PARMOD KUMAR
SANDIP SARKAR

ACADEMIC FOUNDATION
NEW DELHI

www.academicfoundation.com

First published in 2012
by

ACADEMIC FOUNDATION
4772-73 / 23 Bharat Ram Road, (23 Ansari Road),
Darya Ganj, New Delhi - 110 002 (India).
Phones : 23245001 / 02 / 03 / 04.
Fax : +91-11-23245005.
E-mail : books@academicfoundation.com
www.academicfoundation.com

In association with

Institute for Human Development
NIDM Building, IIPA Campus,
I.P. Estate, Mahatma Gandhi Marg,
New Delhi-110 002.
Phone: 23358166, 23321610; Fax: 23765410
E-mail: mail@indindia.org
www.indindia.org

This book is based on the output of the project
Sponsored by Indian Council of Agricultural Research (ICAR)

Cataloging in Publication Data--DK
Courtesy: D.K. Agencies (P) Ltd. <docinfo@dkagencies.com>

Kumar, Parmod.
Economic reforms and small farms: implications for
production, marketing and employment / Parmod Kumar,
Sandip Sarkar.
p. cm.
Includes bibliographical references (p.).
ISBN 13: 9788171889372
ISBN 10: 8171889379

1. Agriculture--Economic aspects--India. 2. Farms, Small–
India. 3. Farm produce--India--Marketing. 4. Rural poor–
Employment--India. 5. Agriculture and state--India. 6. India–
Economic policy--1991- I. Sarkar, Sandip, 1961-, joint
author. II. Title.

DDC 338.10954 23

Contents

List of Tables and Figures

Tables

Figure

Preface

The liberalisation and globalisation strategy initiated in 1991 with manufacturing and trade sectors was later on extended to other sectors including that of agriculture. Withdrawal of movement restrictions, changes in the ECA, removal of licensing requirements and stocking limits, curtailment of selective credit controls and state trading activities and removal of ban on the futures markets of food grains were the major steps initiated by the reform process in agriculture. In the external trade, barring a few exceptions, the exports of all major agricultural commodities are now liberalised. Thus, by now, India has about more than a decade's experience of living under the 'global umbrella'. The problems and stresses of switching over to an open economy are getting known. What are the effects of the reforms implemented so far on various aspects of people's living, especially those of rural areas and what directions the economy is likely to take are the questions that are attempted in this very study. The study brings forth the 'Two Faces of India' and the comparison reveals the weaknesses of the system in two diametrically opposite situations.

The book intends to look at the impact of recent agriculture-related policy changes on the emerging production, marketing and earning status of small *versus* large farmers. Specifically, the study examines: (i) The input-use pattern by small *versus* large farmers to ascertain their degree of technological up-datedness in their production outfits; (ii) The relationships between farm size and intensity of input-use, cropping intensity, yield rates, productivity levels and net returns per hectare cropped for different farm size holdings; (iii) The marketing channels, locale of disposing of marketable surpluses, net effective prices received and compare the proportion of output marketed among small and large farmers; (iv) The degree of on-farm employment diversification, i.e., dairy, fishing, forestry, hunting, etc., in addition to crop production reported by marginal/

small farmers compared with higher groups of farmers; and (v) Quantify the income distribution pattern, in respect of on-farm, off-farm and total household income to ascertain whether the conventional edge of marginal/ small farmers in respect of off-farm employment and earnings is getting lost, especially in areas which are opening up to international production and trade regimes.

In addition to available secondary sources the study uses primary survey conducted on 426 households in Punjab and 430 households in Bihar and a brief Census of 3446 households in Punjab and 3043 households in Bihar. The suggestions for policy are that understanding the current status of globalisation is necessary for setting course for future. Each region (or state) differs in its economic set up and therefore, cannot have a single set of policies to deal with all the states and regions. Whereas agriculture in Punjab set the course of commercialisation way behind in the early seventies with the beginning of green revolution and the state agriculture is now prospecting for a new phase of corporatisation of agriculture leading to the ascendancy of processing and value addition of farm products. On the other hand, agriculture in Bihar still lags behind with small farm orientation led by overwhelming subsistence concerns. The problems that Punjab farmer is facing today relates to the question of sustainability, a consequence of intensive cultivation of land in the state. Bihar farmer on the other hand is trapped in the web of backward infrastructure, low factor intensity resulting in low productivity and consequently low returns from the agriculture. Therefore, different policy actions are *de rigueur* to bring the farmers out of their dilemmas especially the small and marginal ones who are at the ebb of deprivation.

There are large numbers of people who were engaged with the study at various stages. At the outset, Indian Council of Agricultural Research (ICAR) provided the funding for the study and thereby is duly acknowledged. The project proposal was initiated by Prof. G.K. Chadha, the former Vice Chancellor, Jawaharlal Nehru University (JNU) and Chairman SAARC University, who later withdrew from the project because of his pre-occupation with other compulsory involvements. His initial interest in the study and his proposal write-up became the guideline for us later on while conducting the study. His contribution in the book is duly acknowledged. We acknowledge Prof. Manjit Singh of Punjab University, Chandigarh, who was associated with the project from its early stages until the report write-up stage. We also acknowledge, Prof. Sucha Singh Gill, Punjabi University, Patiala, who was also involved with the study at some stages. Prof. Alakh N. Sharma, Director, Institute for Human Development (IHD), remained the source of inspiration throughout the duration of the project. But for his interest in the study, the book would not have taken the present shape. There are numerous people who helped at various stages of the work, especially to mention, Mr. Balwant Singh Mehta, Associate Fellow, IHD, in processing of field survey data and in the generation of tables and Dr. Ramashray Singh, Senior Research Officer, IHD, for taking responsibility of entire field survey of the project. While in the duration of the project we were helped by the field investigators and without their active participation the work would not have been completed. Their contribution is duly acknowledged. In particular we acknowledge Mr. Surinder Goyal and Mr. Joginder Singh, the field survey supervisors in Punjab and Mr. B.K.N. Singh, Mr. Harendra Kumar Choudhury and Mr. Birendra Pathak, the filed survey supervisors in Bihar. The households in Punjab and Bihar not only helped the primary survey team but also provided at times food and shelter during the field work and therefore their support had been immense in completing this study. Finally, Prof. R.S. Deshpande, Director, ISEC and Prof. Sucha Singh Gill, Punjabi University, Patiala, took all the pain in providing blind referee comments on the study that helped immensely in improving the earlier draft of the book. Last but not the least, Ms. Sonali Mukhopadhyay's help in editing the manuscript of the book is duly acknowledged. Not to mention, any errors or omissions are the responsibility of the authors alone.

The book contains seven chapters. Chapter 1 presents the introductory remarks to set the tone with main objectives of the study. Database and methodology of the study are discussed in detail. This is followed by a chapter that is divided into two sections beginning with a section discussing economic reforms in the agriculture sector during the recent years and the general agriculture features covering cropping pattern and performance of agriculture in Punjab and Bihar during the pre and post reform periods. Chapter 3 presents brief findings of the survey data on cropping pattern, productivity, cost, resource use efficiency, production function and farm income structure in the two states under the preview of this study. Production function in Punjab and Bihar are also discussed in this chapter. An analysis of marketed surplus and prices of the selected crops are discussed in the chapter 4. This is followed by a comparative analysis of farm and non-farm employment in Punjab and Bihar in chapter 5. The total earnings from farming and non-farming activities and consumption pattern of food, non-food and other non-durable commodities are presented in the subsequent sections. Chapter 6 summarises the comparison of rural economies of Punjab and Bihar during the pre- and post-reform periods. Last chapter summarises the main findings of the study and puts forward the major areas of policy thrust.

—Parmod Kumar and Sandip Sarkar

1 Portrait of Two Faces

Since July 1991, a new era has commenced in the Indian economy. The development strategy has switched over from inward looking and import substituting to outward looking, open and trade-linked. This strategy has necessitated changes in policy perception, investment priorities and newer inter-sectoral linkages, all geared to gain competitiveness in the external sector. The state is being called upon to play new, facilitating and market-friendly roles. Under the new dispensation, the public institutions, most ostensibly the banking and other corporate financial institutions, have newer, market-driven and 'tougher' priorities to observe. The public and private sector partnership has now to proceed along different lines because, on the one hand, a 'level playing field' has to be ensured to the export-linked domestic producers, and, on the other, the state's interventionist stances have to be selective and market-responsive. In the post-reform years, technology upgradation has become the most inescapable and all encompassing prerequisite for 'staying in the game'. While in certain sectors or activities, technology-on-ground has to be straightened out in terms of the conventional parameters (namely introduction of new products, new inputs, new methods of using inputs, or even new marketing methods and strategies), in others, it is the newer god of 'information technology' that dominates the field. And then, the people who can effectively grapple with the requirements of the new technologies can no more be the uneducated or unskilled workers.

By now, India has about a decade's experience of living under the 'global umbrella'. The problems and stresses of switching over to an open economy are getting familiar. As is typically the case in such transitory phases, in India too, public opinion is not unanimous about the long-run effects of liberalisation. Happily, however, a political consensus is unmistakably discernible throughout the length and breadth of the economy, about the need to liberalise, if India has to take its place in this century as a competitive economic entity. A common opinion is building up about the inevitability of effecting economic reforms to rid the system of the old bureaucratic and other constraints, so as to push the economy to a higher growth trajectory, and about the need to respect private initiatives and capabilities.

That a sort of nationwide political consensus on the need for economic reforms exists is a redeeming feature of the Indian polity. What the effects of the reforms implemented so far on various aspects of people's living are, what directions the economy is likely to take, and how people's working, living and general standard of living would look like after a decade or so, are the questions that are yet to be answered with empirical certainty.

Small and Large Farm Sector in the New Economic Regime

The present study intends to look at the impact of the recent agriculture related policy changes on the emerging production, marketing, employment and earning status of small and large farmers. Marginal and small farmers together constitute a majority of the cultivating households in rural India and are likely to hold a preponderant position in the Indian agriculture setup. The idea is to see how they are faring in all aspects of their economic existence in the post-reform years, in comparison to their counterpart large farmers. There are reasons to believe that marginal and small farmers, as opposed to more propitious large farmers, are likely to face unfavourable circumstances in their farm production and marketing outfits on the one hand, and an unresponsive labour market on the other.

It is also to be seen whether the advantage of labour intensity, which hitherto worked in the favour of small and marginal farmers in compensating their disadvantage of lower operational area, still stands in their favour or is it switching towards large size with increasing mechanisation in agriculture. Therefore, it is essential to see how marginal and small farmers in different parts of India are coping with new production and marketing regimes *vis-à-vis* their counterpart large farmers. It is also to be seen how the various avenues of employment and earnings are getting affected under the open market regimes. This is of special interest in the case of non-farm employment which normally should give small and marginal farmers and landless cultivators more opportunities than their counterpart large farmers who would be absorbed more in the farming sector itself.

Getting Policy Right

The importance and relevance of the study can be seen from many angles. To begin with, marginal and small farms constitute an overwhelming majority of operational holdings in India. What the 'new agriculture' holds for a sizeable majority of Indian farming households should indeed be the first priority on the policy agenda of the nation. In the ultimate analysis, the whole gamut of questions relating to rural development, well-being and poverty must necessarily be seen through the economic well-being of this considerable section of the rural population. No rural development strategy can succeed and no noticeable change could be made in rural poverty unless the weaker sections participate effectively in the production process and register respectable expansion in their employment and earnings. No policy administrator in India can look the other way when a majority of production entities are reported to be losing their ground under the changed policy regimes; the policy administrators must necessarily take cognisance of the institutional, agrarian, marketing and human infirmities that 'small farm producers' usually suffer from, and are alleged to be suffering all the more after the onset of economic reforms. In particular, we sketch out below a very rough description of marginal/ small farmers' problems, in the production, marketing and employment areas, that are believed to have surfaced more glaringly during the recent years. Empirical investigation on a wide scale is extremely essential if indeed the emerging problems are to be scientifically delineated, and relevant policy interventions devised.

Production Effects

There are many elements in the new economic regime that are likely to impinge on the farm production structure of marginal and small farmers in India. First, on the production side, new production technologies (to break the deadlock of the plateau reached by technology in recent years), a new human capital base (for assimilating, absorbing and putting to full use the nuances of new technological outfits) and market-savvy approach of all concerned agencies, are indispensable prerequisites for staying in the emerging competitive economic system. More specifically, different skills are now needed to respond to: (1) promising activities associated with high-value cropping systems; (2) market-oriented crops and more remunerative land-use practices and (3) reductions in production costs for traditional cereal crops (Bathrick, 1998).

It is feared in certain quarters that the reduction in subsidies on fertilisers, electricity, irrigation and credit, needed on grounds of domestic fiscal propriety, is likely to throw marginal and small farmers to a position of disadvantage (besides further exacerbating the inter-regional balances in agricultural growth). Although severely constrained by their limited land base, yet, as the argument goes, until now it has been the subsidised availability of these inputs that has enabled many a marginal and small farmer, especially those in the progressive agricultural regions such as Punjab and Haryana, to keep pace with modern production technology. Until recently, in the matter of per hectare dispersal of institutional credit, per hectare use of fertilisers, the proportion of net sown area under irrigation, the level of cropping intensity, etc., the marginal and small farmers have not shown any significant disadvantage (Vyas, 1998). The future, however, bears great uncertainty in these aspects. On the other hand, medium and large farmers, particularly those in the progressive agricultural regions, are often reported to be 'over using' the subsidised irrigation (especially through tube wells/pump sets), electricity and fertiliser inputs, consequently causing severe problems of groundwater depletion, soil degradation and environmental pollution. Perhaps, it may be desirable to re-orient the input subsidy policy expressly in favour of the small farm producer, which may indirectly serve a larger national interest without compromising on the level of agricultural output. In any case, there is a need to look into the pattern of input use by different categories of farms. The idea is to ascertain the weaknesses of marginal and small farmers when administered price regimes give way to market-determined input price levels.

Second, in the wake of economic liberalisation (e.g. the WTO requirements of intellectual property rights (IPR) either through patents or *sui generis* system), many multinational companies would acquire IPRs, especially in the area of high yielding variety (HYV)/hybrid seeds, and would make Indian farmers buy seeds at high prices. In such a case, the vulnerable position of marginal and small farmers hardly needs to be emphasised.

Third, in days to come, human capital will claim pre-eminence in every aspect (physical, technical or economic) of pre-production, production and post-production decision-making. For example, if and when transgenic crops, already in use in many parts of the world, are introduced on a big scale in India, monitoring of toxicity and allegenicity would be an inescapable requirement not only in the research laboratories and extension outfits, but also at the farm level; a well-informed and scientifically trained farmer alone would be able to ward off the deleterious effects of pest attacks, and capture the full potential of productivity breakthroughs. Again, as is commonly agreed, the competitiveness of the farm sector will depend on the potential for a dramatic reduction in unit cost of production. This can happen either through a shift on the yield/productivity frontier or an increase in input use efficiency. To illustrate from the emerging technology frontiers for wheat, the furrow-irrigated, reduced-tillage bed-planting system (representing an effective combination of wheat planting in beds and traditional ridge-tillage technologies), holds immense promise for making irrigated wheat farming less resource-intensive and more sustainable. Not only that, improvements in fertiliser use (balanced nutrient dosage, observance of timeliness, application of right quantity, etc.), weed control practices and environment-friendly and market-dictated irrigation management/use, all result in marked production efficiencies and cost savings for farmers. But then, these potentials would become a reality far more regularly and far more comprehensively for the educated, better-trained and more-skilled farmers than their brethren who remain bereft of scientific tuning and technical competence. In future, extension would have a far more crucial role to play, however things cannot all be left to the extension outfits; individual initiatives, vision and a progressive outlook would be serious inputs for better farming systems. The role of human capital input would be all the more pervasive and decisive in dry land farming, especially when drought tolerance would have to be combined with heat tolerance, or when nutrient deficiencies are to be dealt side by side with salinity (Pingali and Rajaram, 1998). It is thus high time to take note of the changing complexities of production technology and marketing management at the farm level, and to see if the farming community in general and marginal/small farmers in particular are preparing themselves for meeting the challenges of the 'new Indian agriculture'. The marginal/small farmers' contact with the extension network needs to be looked into more carefully, in the changing context.

Fourth, ever since the introduction of economic reforms in 1991, a view has emerged that there is a tremendous scope of reaping economies of scale in farming, especially in respect of tradable agricultural commodities. The implication is that marginal and small farms must give way to large scale farming, through relaxation of ceilings on landholdings and tenancy laws. It is argued that marginal and small farmers produce mainly for self-consumption, and they have very little marketable surplus to offer. The income gains to domestic agriculture, in the event of international price increases following the withdrawal of subsidies and other supports by the developed western economies, would accrue if and only if production is organised on a large scale. A case is thus made for relaxation of land ceiling and tenancy laws, which might possibly encourage both large scale ownership cultivation as well as corporate farming.

As a matter of fact, some states have already gone ahead with their ceiling relaxation agenda and it is feared that land is gradually moving away from marginal and small farmers (Pandiyan, 1996; Reddy, 1997). In some other areas, reverse tenancy is reported to have risen rather significantly during the 1990s (Vyas, 1998; Kumar, 2005). Although, the incidence of such land transfers may be limited and confined to specific regions, yet the future of small farming and small landowners hangs in balance. The issue assumes crucial significance inasmuch as the non-farm avenues of employment are not likely to come easily to such households, in the market-driven and competitive labour market of the future. In fact, there is no reason that a small producer in India cannot benefit from economic liberalisation when strong institutional support is available.

Fifth, it is reported that in many areas, reverse tenancy has witnessed a steady expansion in the recent past, largely prompted by considerations of economies of scale, price efficiency and marketing competitiveness, both in the domestic and international markets. If that is so, the marginal and small farms must be getting edged out of the land lease market, or facing more exacting terms and conditions of the land lease contracts, or may be more variously involved in factor market interlocking, and so on. There is no empirical evidence about all these developments for the recent past, particularly for the nineties. It is time, therefore, to ascertain the changing contours of land lease market from the point of view of the marginal and small farmers' access to land, and the future viability of their farm production enterprises.

Finally, it is reported that in certain regions, the areas that are witnessing fast transition to commercial and export-linked farming and where considerations of economies of scale are gaining heightened significance, marginal and small farmers are opting out of cultivation (Kumar, 2006). It is high time to ascertain if these fears exist on the ground. If that is the case, it is equally important to know what are viable economic alternatives for the marginal and small farmers, constituting as they do a vital majority of the cultivating households in each region. In particular, it would be instructive to know what the next generation of workers from the present-day farming households visualises as their economic path.

Marketing Problems

New paradigms are emerging on the marketing side as well. First, marketing of farm produce would now be a different and complex affair. The era of 'produce and then sell' has ended. Knowledge of consumer needs and product promotion has become paramount for linking local capacities with national, regional and international needs. Market information services and intelligence systems, especially those providing information about new consumers and emerging product demands, competing prices, changing quality and health standards (such as sanitary and phytosanitary standards in the context of trade) will become increasingly important to the changing agriculture. How far the required market intelligence travels to the hitherto subsistence-oriented marginal and small farmers is a moot question. One thing quite certain however is that the future farmer of India cannot be a reincarnation of the parents. The modern farmers will need new skills, new perceptions, new commercial understandings and new economic calculus if they are to do well in marketing, just as they have to become more efficient and well-informed producers. In a general sense, educated farmers of tomorrow will hold the ground well, just as their uneducated or illiterate neighbour will find it increasingly difficult to grapple with the complex production and marketing issues. The vulnerable position of marginal and small farmers in this respect is perhaps too obvious to be emphasised.

Second, at present, India is competitive only in respect of a few specific agricultural commodities (Nayyar and Sen, 1994). There is tremendous scope, however, for export of horticultural and dairy products. For example, in the recent plan documents, eight fruits (i.e., mango, grape, banana, sapota, pomegranate, apple, citrus and litchi) have been identified for thrust on exports. It is widely felt that small farmers too can cultivate horticultural crops if only adequate institutional support and extension advice are forthcoming. For example, some recent studies show that contract farming under which small farmers produce fruit and vegetables for cooperative or corporate bodies against specified contracts (such as soyabean farmers of Rajasthan, potato farmers of Punjab) can enable marginal and small farmers to chip off a share in corporate gains. In general, it is true nonetheless that lack of irrigation and soil moisture, high quality planting material, processing units, infrastructural support for packaging, storage, cooling and transport would stand in the way of small farm diversification. The question of public investment thus assumes significance from a marketing angle too, as it is highly crucial for improvement of the small farmers' production base. The profile of public investment in recent years poses some disturbing signals: The abysmal record of some states in this regard forebodes distressing times ahead for marginal and small farmers in those areas, if we bear in mind the public and private sector investment complementarities especially at the lower end of the farm size ladder. It is time, therefore, to look into the level and pattern of investment by marginal and small farmers in the backward regions in contrast to medium and large farmers, and to those in the progressive farming regions.

Employment and Earning Effects

Significant changes are reported to be taking place in the level and pattern of marginal and small farmers' employment and earnings. As is well known, a sizeable proportion of marginal and small farmers seek employment outside their farm, besides engaging themselves in work on their own farms as cultivators, including related work on dairying, fishing, forestry, hunting, etc. Outside employment may be as a wage labourer in agriculture, or in some non-farm activity. The range of non-farm activities is wide and encompasses those self-employed in industry, construction, transport, trade and other services, as well as those employed by others in similar occupations as wage-paid workers. Seasonal and contractual jobs not connected with farming as such, available locally in the village or in the nearby semi-urban or urban-industrial-commercial towns, are also a part of their non-farm employment.

It is thus clear that employment and earning effects of economic reforms would emanate not only from the technological changes including new methods of carrying on field crop operations and cropping pattern adjustments occurring on their own farms, but also from the whole network of outside activities with which the small/marginal farmers are associated. For example, the introduction of a new short duration crop or a new variety of a crop may reduce man-days of labour input for that crop, but may increase the total self-employment time through increased level of cropping intensity. Similarly, if a crop operation is accomplished through hired machine services, the operation-specific level of self-employment may decline but compensatory additional employment may be forthcoming because a larger volume of output is to be harvested, marketed or semi-processed. The same tendency may operate on other farms as well, and may affect wage employment of our marginal and small farmers both positively and negatively, and the net overall position needs to be looked into. It is quite reasonable to presume nonetheless that agricultural wage-employment and wage rates for marginal and small farmers' employment on big farms will get affected, since the latter are sure to effect labour-saving technological changes on the one hand, and face unsubsidised input regimes affecting their cost calculus, on the other. To cap everything else, 'the new agriculture' itself would be highly skill and education intensive, and wage-paid employment even within agriculture, most strikingly among big farms linked with the export market, is bound to leave out a majority of job aspirants coming from marginal and small farming households: Their low educational achievements, even 10-15 years hence, would be their Achilles' heel (Chadha, 1999).

Employment prospects outside agriculture are more difficult to predict. In the medium range, regular wage-paid non-farm employment may decline; self-employment may not decline sharply while casual wage labour may increase sizeably. All these developments will affect the total non-farm employment, wages and earnings of marginal and small farmers in a big way. This, in turn, would adversely affect their prospects for crossing over the poverty line. There are numerous studies to show that non-farm employment and earnings make a decisive dent into the poverty ratio among the lower strata of the rural households (Chadha, 1994). In the longer run, casualisation may grow apace, and self-employment may remain accessible to that declining segment of marginal and small farming households who are able to improve their skills and educational standards. The worst affected are feared to be the female job aspirants.

Setting the Parameters

The study aims to analyse different aspects, attributes and problems concerning production, marketing, employment and earnings of small and large farmers in the light of ongoing changes occurring in the global and domestic arena. The main objectives of the study are outlined in the following paragraphs.

Production

To take stock of farm and non-farm production assets and see the level and pattern of investment by marginal/small farmers in comparison to medium/large farms to ascertain if marginal/small farmers are facing institutional constraints in raising investment resources. Further, the study aims to find out the nature and severity of agrarian constraints being faced by marginal/small farmers. The question whether land is gradually drifting away from such farmers and whether a heightened degree of reverse tenancy has set in those regions, specifically whether small/marginal farmers are opting out of cultivation, is to be empirically tested.

To examine the nature of cropping pattern diversification, if any, being followed by small *versus* large farmers, especially in regard to the substitution of high value commercial crops for food crops, and to examine whether a drift is taking place towards horticulture, floriculture, fisheries etc. The study will also examine the input use pattern by marginal/small farmers, compared with higher categories of farm operators, to ascertain the degree of technological updatedness in their production outfits. Further, relationships between farm size and irrigation base, intensity of input-use including labour use, cropping intensity, yield rates, productivity levels, net returns per hectare cropped, and earnings per day and per worker would be examined.

Marketing

To examine the marketing channels and time pattern of sales, locale of disposing of marketable surpluses, net effective prices received, etc., by marginal/small farmers compared with medium and large ones. The study will also compare the proportion of output marketed among small and large farmers to see if any kind of distress sale is prevalent among the small and marginal farmers.

Employment and Earnings

To examine the on-farm employment sources and man-days of work put therein by marginal/small farmers in relation to medium/large farmers, in particular, the degree of on-farm employment diversification (dairying, fishing, forestry, hunting, etc., in addition to crop production) reported by marginal/small farmers compared with higher groups of farmers. Further, the study will probe total earnings from each type and the total of on-farm employment, to see if marginal/small farmers fare poorly in relation to medium/large farms.

To examine the off-farm employment sources and man-days put therein by marginal/small farming households in relation to medium/large farmers and in particular, to examine whether an inverse relation exists between farm size and the proportion of work time spent on off-farm employment. To find total earnings from each type and aggregate of off-farm employment and to examine whether marginal/small farmers fare poorly in relation to medium/large farmers. In particular, the qualitative differences (e.g. investment in capital-intensive mechanical transport equipment by medium/large farmers against labour-intensive human-driven equipment by marginal/small farmers) along the farm size continuum needs to be brought out to test the hourglass hypothesis.

To quantify the income distribution pattern in respect of on-farm, off-farm and total household income, to ascertain whether the conventional edge of marginal/small farmers in respect of off-farm employment and earnings is getting lost, especially in areas which are opening up to international production and trade regimes. Further, the study will look into the composition of employment (say, self against wage employment) along the farm size ladder. In particular, it will see if the proportion of casual on- and off-farm wage employment is significantly higher among marginal/small farming households than among the higher farm size groups.

Choice of Regions

With much thought, two regions were selected for this study, namely Punjab and Bihar. These two states represent the most progressive and most backward agriculture regions in the country respectively. The logic for opting for these two states is defined in the following section.

The Punjab study would reveal whether the ground gained by marginal and small farmers earlier under the green revolution umbrella is getting lost under 'new agriculture'. It will also reveal whether the new technological outfits or marketing methods are beyond their reach even though the basic ingredients of a strong production base (i.e., irrigation, consolidated holdings, extension network, assured product prices, assertive agrarianism, etc.) are still in position. While it is true that during the nineties, Punjab has not shown any significant switchover from the dominant wheat-rice production cycle inherited since the seventies, the expanding commercial status of agriculture in the state, which has now transcended the national frontiers, would still highly justify the inclusion of Punjab in our study.

Bihar's agriculture, on the other hand, reflects a total contrast to Punjab; the state still displays the results of a combination of a caste-ridden rural society, exploitative agrarian outfits, lack of rural infrastructure, and gross negligence by the state apparatus towards agricultural development and rural well-being. Agriculture in Bihar still operates on the lines of subsistence farming, in complete contrast to the ascendancy of commercial farming in Punjab.

Therefore, the survey results in these two regions would throw up tremendous divergences across these two regions and across various farm size categories. In plain terms, a cross-section of inter-regional and inter-farm size comparisons is bound to throw up meaningful production-marketing-employment typologies which can be drawn upon for more effective policy interventions.

The Scaffolding

Given the objectives of the study, a detailed analysis was undertaken based on both secondary and primary data sources. For the secondary data we used Government of India reports and publications like the *Economic Survey*, Ministry of Finance, Directorate of Economics and Statistics; *Statistical Abstracts of India* and of various states; *Outline of India* and of various states, Department of Economics and Statistics and Central Statistical Organisation, Government of India.

Following this, the present study is based on a large database collected from two rounds of primary survey undertaken in Bihar and Punjab in the year 2003-04. We covered 16 villages in Punjab and 12 villages in Bihar, for conducting the primary survey. A comparable design of survey for Punjab and Bihar was utilised, whereby all the districts of the two states were divided into four zones, depending on agro-climatic conditions, cropping pattern and development of agriculture. From each zone, one district was selected randomly for the survey. Further, two blocks were selected from each of the districts, covering both the extremes of development in the district. At least one village was selected from each block, in such a way that it represents the focus of the main objectives of the study. In Punjab, two villages each were chosen from every block. In Bihar, we chose two villages from a relatively advanced block and one village from an underdeveloped block. The change in sample design in the case of Bihar was deliberate, based on the orientation of the study. The idea was that since level of infrastructural and agricultural development is considerably low in Bihar, compared to that of Punjab, it would be better to give additional emphasis to more developed districts, to ascertain the possible impact of WTO on agriculture.

Since our study is related to different categories of farmers from large to marginal, we have followed the method of stratified probability proportion random sampling. With this design, we selected a sample of 400 to 450 households from each state for the purpose of detailed study.

The study involved household survey in two rounds and community level data. The household data was based on a sample survey conducted in 12 villages in Bihar and 16 villages in Punjab. The entire fieldwork was carried on in two rounds in the years 2003-04. The first round of the survey was conducted in two parts. First, a listing of all the households in the village was undertaken along with a one page census questionnaire, to get to know basic household characteristics, land size, broad economic activity undertaken by households, particularly in agricultural and allied activities, and other agricultural and livestock assets. In all, 3,446 households in Punjab and 3,043 households in Bihar were covered in the Census. The questionnaire was specifically designed to permit classification of households into marginal, small, medium and large landowners.

On the basis of the Census questionnaire, households were categorised into landless, marginal, small, medium and large landholding categories, and from each strata, a certain number of households were randomly chosen so that distribution of households in different landholding categories corresponded to district landholding patterns in both states. In this fashion as a whole, we selected 426 households in Punjab and 430 households in Bihar. The breakdown of the number of households district-wise is presented in Table 1.1.

The second round of the survey was undertaken to capture the details of various seasonal activities that could not be effectively captured through the earlier information collected through the single round survey. These activities were cultivation, casual work and non-agricultural self-employment, which are casual in nature. Information regarding these activities was collected six months after the first round for sample households.

In addition to these data, community level data covering 28 villages in two states were collected to facilitate inter-village comparisons. It was initially collected with the first round survey, but the gaps were filled up in the subsequent round of household survey. Village schedules covered a large segment of the area under household survey. In all the 28 villages for which a detailed household survey has been conducted, employment, wage rate and similar data were more accurate than corresponding data from the community level survey.

Table 1.1

Distribution of Sample Households across Various Farm Sizes based on Owned Land

(Number of households)

	Marginal	*Small*	*Medium*	*Large*	*Landless*	*Total*
Punjab						
Amritsar	11	9	13	5	13	51
Bathinda	7	4	3	16	27	57
Hoshiarpur	15	11	6	1	19	52
Jalandhar	14	9	10	3	17	53
Mansa	11	11	7	8	15	52
Moga	11	9	11	9	18	58
Patiala	11	5	13	7	15	51
Sangrur	10	6	8	11	17	52
Total	90	64	71	60	141	426
Bihar						
Bhojpur	39	17	18	9	26	109
Gopalganj	55	22	12	0	18	107
Jahanabad	39	16	20	7	26	108
Khagaria	35	19	12	13	27	106
Total	168	74	62	29	97	430

In all the villages covered in the survey, some information was obtained from the Census questionnaire like population characteristics, main occupations, livestock activities, and diversified agricultural activities. In addition, some information on infrastructural facilities as well as on land use patterns was available in the village revenue records. Local officials maintain detailed information pertaining to the land use and cropping pattern. However, we found that written official documents at this level are quite unreliable and Census data are difficult for use, since many villages display erratic fluctuations from one census to the other. Price data were obtained from local markets and school enrollment data were collected from school records.

The main source of community data was participants from within villages. These included village officials, groups of cultivators and agricultural and non-agricultural labourers, school teachers and various others, ranging from politicians to local literary figures. Information on wages was obtained from both employers and wage labourers, and about tenancy from landlords and tenant households. Qualitative information was also collected form various respondents. To reduce the element of bias in each case, care was taken to collect information from informed groups, (say for cropping patterns from a group of cultivators, and for school enrollment from a group of teachers).

Sample Size and Composition

The farm size categories are defined below:

1. Marginal farms - with operated area up to 1.00 hectare
2. Small farms - with operated area between 1.01 and 2.00 hectares
3. Medium farms - with operated area between 2.01 and 4.00 hectares
4. Large farms - with operated area more than 4.00 hectares
5. Landless - those who did not operate any land during the reference year.

Overview

The study is presented in seven chapters. Chapter 1 presents the introductory remarks to set the tone and introduces the main objectives of the study. The database and methodology of the study are discussed in detail. Chapter 2 is divided into two sections, beginning with a section discussing economic reforms in the agriculture sector during the recent years, and the general agriculture features covering cropping pattern and performance of agriculture in Punjab and Bihar during the pre- and post-reform periods. This is followed by a brief analysis of Census data collected through a two-page questionnaire, covering about 3,000 households in Punjab and Bihar each.

Chapter 3 presents brief findings of the survey data on cropping pattern, productivity, cost, resource use efficiency and farm income structure in the two states under study. Findings on the production function in Punjab and Bihar are also discussed in this chapter.

An analysis of marketed surplus and prices of the selected crops are discussed in the chapter 4. Here, the Section I presents the marketed surplus by farm size at the aggregate level for the major crops. Section II makes a comparison of prices obtained by farmers in the two states. Section III presents the marketing channels used by the farmers in disposing off their surplus, followed by Section IV, which analyses the cost of marketing among the selected households. Section V presents a detailed regression analysis to find out the main determinants of marketed surplus and prices received by the farmers.

Chapter 5 presents a comparative analysis of farm and non-farm employment of Punjab and Bihar. The total earnings from farming and non-farming activities and the consumption pattern of food, non-food and other non-durable commodities are presented in the subsequent sections.

Chapter 6 summarises the comparison of the rural economies of Punjab and Bihar during the pre- and post-reform periods. The chapter begins with a comparison of social indicators of the small and large farmers during these two periods. Section II presents land transactions during the pre and post-reform period, followed by the Section III presenting the asset holdings by the households. Section IV makes a comparison of income and input use by small and large farmers during the two periods. Section V presents the major problems and difficulties faced by the selected households during the current phase.

Chapter seven summarises the main findings of the study and puts forward the major thrust areas for policy.

2 Reforms and the Agricultural Economy of Punjab and Bihar

Meso-Level View

The Indian economy remains a predominantly agrarian economy although the contribution of the primary sector to GDP has declined considerably. More than 50 per cent of the population is still dependent on agriculture, while its share in GDP has come down to around 15 per cent. Thus, over a period of five decades, the shift in the labour force from agriculture to non-agriculture activities was less than 20 per cent, whereas GDP change was about 40 per cent. Low labour productivity continued to be the major limitation of Indian agriculture. More than 40 per cent of the agricultural workforce consists of landless agricultural labourers. In view of the large size of the rural population and agricultural workforce, the extent of rural poverty and unemployment, and the ever-rising rural-urban gap, agricultural growth and gains in agricultural productivity need special attention and treatment (Ghuman, 2002: 125). As the manufacturing and service sectors still depend on the performance of the agriculture sector, the latter still holds the key to ensure pro-poor growth, provide food security, reduce poverty and generate employment opportunities for the rural as well as urban masses.

India is not a large agricultural trader in the world market.[1] Until 1991, India followed an inward-looking development strategy with a trade regime characterised by quantitative restrictions, licensing and high tariffs. As a result, domestic markets were virtually insulated from changes in world market prices. The Government intervened heavily in both product and input markets, through price support programmes backed by government procurement and input subsidies. These interventions resulted in net taxation

1. Agricultural imports and exports were of the order of US $3.6 billion and $7.9 billion respectively, in 2003-04 and accounted for 4.6 per cent and 12.4 per cent respectively, of total imports and exports.

of the agricultural sector, while the non-agricultural sector received protection. The extent of the total taxation of the sector was estimated at 29 per cent of the value of agricultural production during 1971-1985, 18 per cent during 1986-1991, but only 9 per cent during 1992-95 (Pursell, 1999: 30).

India has undertaken a substantial degree of trade liberalisation, beginning in 1991. However, initially the focus remained on manufacturing including capital goods. Liberalisation was extended to agriculture in 1994, when the Government lifted a number of restrictions on imports and exports, simplified trade measures, and reduced public interventions in domestic markets. Major emphasis was laid on increasing exports of agriculture-based products. To reform the domestic market sector, movement restrictions were lifted across and within states on food grains, and futures trading was allowed on certain agricultural commodities. Similarly, steps have been taken for the development of agro-processing and cold storage facilities, and changes have been made for simplification of licensing and stocking limits for the wholesale and retail trade for certain agricultural commodities (NCAER, 2001: 190). There are also indications of corporate interest in some sub-sectors of agriculture. However, a comprehensive agenda for reforms in agriculture has yet to emerge (Thamarajakshi, 1999; Rao, 2003; Chand, 2004).

As a result of trade and exchange rate liberalisation and significant reduction in the protection to the domestic industry, the incentive framework for Indian agriculture has improved considerably (Rao, 2003: 615). The terms of trade for agriculture have improved and there has been a record rise in agricultural exports. Private investment has increased at a rapid rate in the post-reform period. The full benefits of macro-economic reforms on agriculture are expected to materialise in due course of time. Against this background, we have discussed here the general features of agriculture that include cropping pattern and the performance of agriculture in Punjab and Bihar during the reform period.

Agricultural Economy at Two Extremes: Bihar and Punjab

The Bold Features: Production Sector

A comparative analysis of Punjab and Bihar's agriculture presents quite a contrasting picture. Whereas Punjab has remained the granary of

India since the Green Revolution in the sixties and seventies, Bihar lags far behind in the adoption of new technology, productivity and rural infrastructure. While agriculture remains the prime activity and provides employment to more than two-thirds of the population in both the states, the importance of agriculture in gross production is falling at a faster rate in Punjab as compared to Bihar. The share of agriculture in gross domestic product (GSDP) declined from 47 per cent in 1960-61 to 26 per cent in 1998-99 in Punjab, while it came down at a much slower pace in Bihar from 45.8 per cent in 1980-81 to 37.6 per cent in 2003-04. Further, the rate of transformation of Bihar's economy in terms of growing importance of non-farm and non-agricultural sector is much slower compared to Punjab and other states. Lack of industrialisation and poor rural-urban growth linkages have further worsened the employment position in Bihar, so the percentage of population dependent on agriculture or disguisedly underemployed in agriculture is relatively higher in Bihar compared to Punjab. It is also notable that whereas Punjab is making strides in diversifying its agriculture towards more processing and value addition and dairy related activities (Johl, 2002), the agricultural economy of Bihar is almost entirely based on primary output. There is very little agro-processing and even primary processing like milling, oil extraction etc., are based on primitive and high-cost technologies. Before reorganisation, Bihar was contributing 13 per cent of fruit and vegetable production in the country but accounted for only 1.28 per cent of processing units (Roy, 1997).

A large share of cultivable area still remains unused in Bihar, whereas almost all the cultivable land is under the plough in Punjab. Eighty-five per cent of the reported area was cultivated in Punjab compared to 59 per cent in Bihar (Table 2.1). However, 24 per cent of reported area was under forests and non-agricultural uses in Bihar, while Punjab had only 13 per cent reported area under such uses. The cultivable area that could be brought under the plough or was liable to be cultivated belonged to the fallow land that made a huge difference in net cultivated area in these two states. There was hardly any fallow land (less than one per cent) in Punjab, while more than 8 per cent of the reported area was left fallow in Bihar. Area under forests had increased slightly in both the states. On the other hand, pastures and marginal lands (cultivable and uncultivable wasteland

Table 2.1

Comparative Land Use in Punjab and Bihar

('000 hectares)

Classification of Land	Punjab				Bihar			
	TE 1981-82	TE 1997-98	TE 2000-01	TE 2008-09	TE 1981-82	TE 1997-98	TE 2000-01	TE 2006-07**
Reporting area	5033 (100)	5033 (100)	5033 (100)	5033 (100)	17330 (100)	17330 (100)	17330 (100)	9366 (100)
Forests	217 (4.3)	295 (5.9)	305 (6.1)	268 (5.3)	2826 (16.3)	2949 (17.0)	2949 (17.0)	622 (6.6)
Area under non-agricultural uses	434 (8.6)	378 (7.5)	330 (6.6)	484 (9.6)	1723 (9.9)	2372 (13.7)	2429 (14.0)	1647 (17.6)
Barren and uncultivable land	96 (1.9)	78 (1.6)	56 (1.1)	25 (0.5)	1005 (5.8)	1010 (5.8)	1010 (5.8)	436 (4.7)
Total land not available for cultivation	530 (10.5)	456 (9.1)	385 (7.7)	509 (10.1)	2728 (15.7)	3382 (19.5)	3439 (19.8)	2083 (22.3)
Permanent pastures & other grazing land	4 (0.1)	5 (0.1)	6 (0.1)	3 (0.1)	141 (0.8)	109 (0.6)	105 (0.6)	17 (0.2)
Land under miscellaneous tree crops & groves	4 (0.1)	10 (0.2)	8 (0.2)	4 (0.1)	213 (1.2)	339 (2.0)	344 (2.0)	240 (2.6)
Cultivable wasteland	48 (1.0)	28 (0.6)	29 (0.6)	10 (0.2)	452 (2.6)	349 (2.0)	322 (1.9)	46 (0.5)
Fallow lands, other than current fallows	0 (0.0)	10 (0.2)	4 (0.1)	1 (0.0)	950 (5.5)	955 (5.5)	923 (5.3)	130 (1.4)
Current fallows	44 (0.9)	57 (1.1)	40 (0.8)	38 (0.8)	1960 (11.3)	1889 (10.9)	1812 (10.5)	660 (7.1)
Total fallow land	44 (0.9)	66 (1.3)	44 (0.9)	39 (16.8)	2910 (0.8)	2845 (16.4)	2736 (15.8)	791 (8.5)
Net sown area (NSA)	4194 (83.3)	4172 (82.9)	4255 (84.5)	4182 (83.1)	8026 (46.3)	7356 (42.4)	7435 (42.9)	5561 (59.4)
Gross cropped area (GCA)*	6742 (134.0)	7813 (155.2)	8288 (164.7)	7881 (156.6)	10730 (61.9)	10059 (58.0)	10027 (57.9)	7459 (79.7)
Area sown more than once	2548 (50.6)	3641 (72.3)	4033 (80.1)	3700 (73.5)	2704 (15.6)	2702 (15.6)	2592 (15.0)	1898 (20.3)
Net irrigated area	3438 (82.0)	3899 (93.5)	3736 (87.8)	4088 (81.2)	2906 (36.2)	3604 (49.0)	3644 (49.0)	N.A.
Gross irrigated area	5816 (115.6)	7414 (147.3)	7854 (156.1)	7693 (152.9)	3533 (32.9)	4608 (45.8)	4789 (47.8)	N.A.
Cropping intensity (GSA/NSA)	1.61	1.87	1.95	1.88	1.34	1.37	1.35	1.34

Notes: i) Bihar includes the area of Jharkhand.

ii) The figures in parentheses are respective percentages of the total reported area except for the last two rows which represent percentage of net sown area and percentage of gross cropped area, respectively.

* % gross cropped area exceeds 100 because of area sown more than once.

** The area for TE 2006-07 is for Bihar after division of the state into Bihar and Jharkhand while previous years figures refer to undivided Bihar.

T.E. Triennium ending.

Source: Land Use Classification and Irrigated Area (1997-98 & 1998-99), Ministry of Agriculture, Government of India.

under village commons), and river and tank beds had also been brought under cultivation. Vertical intensification was high in both the states, as cropping intensity was measured at 1.35 per cent in Bihar, which compares well with 1.95 per cent in Punjab, given the fact that 95 per cent of gross cropped area was irrigated in Punjab as against 48 per cent in Bihar.

The contrast in distribution of operational landholding and area by land size is presented in Table 2.2. The average size of landholding in Punjab was 3.94 hectares in 2005-06. In Bihar, average size of landholding

Table 2.2

Distribution of Operational Holdings in Punjab and Bihar

	Marginal	*Small*	*Semi-Medium*	*Medium*	*Large*	*Marginal plus Small*	*Average Size (Hectares)*
Punjab							
Percentage of Holdings							
1980-81	19.21	19.41	27.98	26.20	7.20	38.66	3.79
1985-86	23.54	19.08	26.72	23.90	6.76	42.61	3.77
1990-91	26.50	18.26	25.87	23.37	6.00	44.76	3.61
1995-96	18.66	16.74	29.28	28.00	7.32	35.41	3.79
2005-06	13.45	18.23	31.87	29.46	7.10	31.69	3.94
Percentage of Area							
1980-81	-	-	20.33	40.26	29.18	10.26	
1985-86	-	-	20.47	38.72	29.85	10.96	
1990-91	4.07	8.13	20.88	40.22	26.70	12.20	
1995-96	2.94	5.79	20.09	42.30	28.89	8.73	
2005-06	2.09	6.51	21.55	42.92	26.92	8.60	
Bihar							
Percentage of Holdings							
1980-81	75.44	11.28	8.47	4.20	0.61	86.72	0.99
1985-86	-	-	8.12	3.45	0.44	87.98	0.93
1990-91	78.61	11.09	7.29	2.71	0.30	89.70	0.83
1995-96	80.15	10.78	6.65	2.22	0.20	90.92	0.75
2005-06	89.64	6.67	2.99	0.68	0.01	96.32	0.43
Percentage of Area							
1980-81	26.67	14.89	23.44	24.48	10.53	41.55	
1985-86	-	-	23.80	21.04	7.70	47.46	
1990-91	33.43	18.19	23.98	18.45	5.96	51.62	
1995-96	36.24	18.89	24.02	16.37	4.48	55.13	
2005-06	53.00	19.58	18.16	8.73	0.54	72.58	

Source: Agricultural Census Division, Ministry of Agriculture, New Delhi.

was much smaller: It was only 0.43 hectares during the same year. Over time, landholdings got fragmented in Bihar in consonance with the trend prevalent at the all-India level as well as in many states. Conversely in Punjab, however, a trend of consolidation of holdings was visible, thanks to the prevalence of the phenomenon of 'reverse tenancy', also observed by a number of recent studies (Singh, 2004: 5583; Kumar, 2005: 3). The average size of holdings was falling till the early nineties but increased thereafter. It is apparent from the data presented in the table that the number of marginal holdings started declining in the early nineties in Punjab, unlike Bihar, where the number of marginal holdings and their area was continuously increasing. Due to the high commercialisation and mechanisation of agriculture, and due to increasing cost of cultivation, the marginal holdings in Punjab are becoming unviable, causing marginal farmers to switch out of agri-business in favour of either wage earning or other formal or informal employment (Kumar, 2005: 3).

It is the predominance of marginal holding that has brought down Bihar's average size of holdings. Almost 90 per cent of all holdings in Bihar were marginal in contrast to only 13 per cent in the case of Punjab. The curve for distribution of operational landholdings as well as trends in the area were downward sloping in Bihar, while both were upward sloping in the middle in the case of Punjab (Figure 2.1). The upward slope of the curve in the middle in Punjab indicates an increase in the number and area of holdings for the semi-medium and medium farmers who accounted for around 60 per cent of area as well as holdings in the state. Only the small land size group had the most comparable share of land area and holdings both in Punjab and Bihar. One major factor that strengthened the preponderance of marginal holding in land ownership pattern in Bihar was the pattern of leasing-in land. For marginal landholders, the share of leasing-in land to total leased-in land was very high in Bihar: It accounted for 78 per cent as opposed to hardly 29 per cent in the case of Punjab (Table 2.3). It can also be seen from the fact that 85 per cent of all leased-in area in Bihar was accounted by marginal landholders as opposed to 24 per cent in the case of Punjab. In Punjab, the phenomenon of reverse tenancy was quite prominent. Medium and large landowners leased-in 36 per cent of all leased-in area, as opposed to a meager 0.36 per cent in the case of Bihar.

Figure 2.1

Distribution of Operational Holdings in Punjab and Bihar (1995-96)

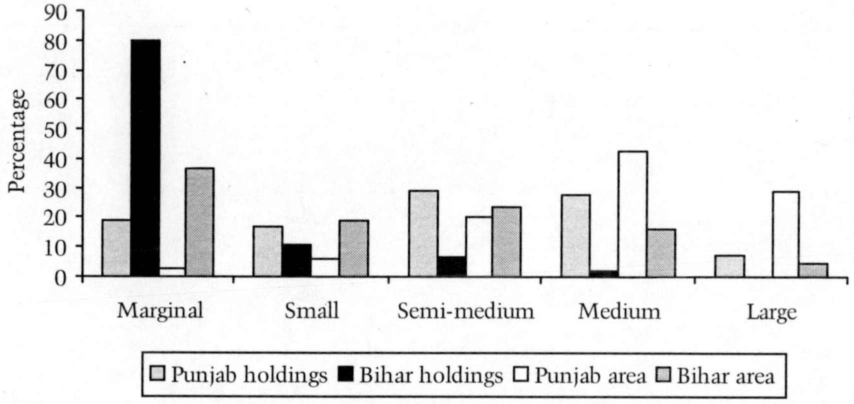

Table 2.3

Leased-in Area by Land Size Holdings in Punjab and Bihar (1991-92)

Land Holdings	*Percentage to Total Leased-in Area by the Households*		*Percentage of Leasing-in to Total Leasing-in Land by the Households*	
	Bihar	*Punjab*	*Bihar*	*Punjab*
Marginal	85.06	24.22	77.94	29.23
Small	9.96	18.16	16.55	12.9
Semi-medium	4.09	26.72	5.15	22.36
Medium	0.89	29.48	0.36	35.52
Large	0.00	1.37		

Source: Srivastava 2000.

The cropping pattern was not well diversified in both Bihar and Punjab. Food grain crops dominated in both these states. Among the food grains, it was rice-wheat rotation that dominated the cropping cycle since the late eighties or early nineties in both the states. These two crops accounted for around 79 per cent of gross cropped area in Punjab and around 70 per cent in Bihar. In Punjab, apart from rice and wheat, cotton was the only other important crop, occupying 7 per cent of gross cropped area. In Bihar, maize was an important cereal crop with a share of 7.7 per cent in its gross cropped area, while *kharif* and *rabi* pulses accounted for 7 per cent of gross cropped area in the state. In Punjab, the share of other

Table 2.4(a)

Cropping Pattern in Punjab

(Percentage to GCA)

Crop	1950-51	1960-61	1970-71	1980-81	1990-91	1995-96	2001-02	2008-09
Paddy	2.9	4.8	6.9	17.5	26.9	28.0	30.2	34.6
Wheat	27.3	29.6	40.5	41.6	43.6	41.8	41.5	44.6
Maize	6.3	6.9	9.8	5.6	2.5	2.2	2.0	1.9
Bajra	*5.2*	*2.6*	*3.7*	*1.0*	*0.2*	*0.1*	*0.1*	*0.1*
Barley	2.4	1.4	1.0	0.9	0.5	0.5	0.3	0.2
Total coarse cereals	13.9	11.2	14.5	7.6	3.2	6.7	2.4	2.2
Total pulses	23.8	19.1	7.3	5.0	1.9	1.3	0.7	0.3
Total food grains	67.9	64.7	69.2	71.7	75.6	77.8	74.8	81.6
Total oil seeds	3.3	3.9	5.2	3.7	1.3	1.9	1.1	0.8
Sugarcane	2.2	2.8	2.3	1.0	1.3	1.7	1.7	0.1
Cotton	5.4	9.4	7.0	9.6	9.3	9.7	7.4	6.7
Total vegetables	1.2	1.2	0.9	1.1	0.7	1.0	1.3	1.5
Total fruits	0.8	0.6	0.6	0.4	0.8	1.1	0.5	0.8
Other crops	19.2	17.7	14.8	12.6	11.0	10.7	13.2	7.7

Source: *Statistical Abstracts of Punjab* (various years).

Table 2.4(b)

*Cropping Pattern in Bihar**

(Percentage to GCA)

	1980-81	1990-91	1995-96	1998-99	2001-02	2003-04*	2007-08*
Paddy	49.79	51.41	50.27	50.73	50.26	44.37	44.67
Wheat	15.74	18.74	21.30	20.88	21.85	24.82	26.08
Maize	7.92	6.34	7.16	6.82	6.84	7.51	7.73
Barley	1.07	0.53	0.44	0.42	0.34	-	-
Total coarse cereals	12.13	8.59	8.99	8.44	8.23	8.40	8.33
Gram	1.76	1.60	1.41	1.26	0.68	-	0.85
Arhar	0.84	0.63	0.71	0.66	0.65	-	0.36
Other pulses	9.67	8.98	7.08	5.62	6.84	-	-
Total pulses	12.27	11.21	9.20	7.53	8.17	9.09	7.36
Total food grains	*89.93*	*89.95*	*89.76*	*87.58*	*88.51*	*86.68*	*86.44*
Total oil seeds	2.29	2.29	2.16	2.10	1.98	2.10	1.69
Jute & *mesta*	1.79	1.62	1.41	1.60	1.58	2.05	1.81
Sugarcane	0.99	1.42	1.25	1.07	1.24	1.28	1.33
Tobacco	0.11	0.15	0.18	0.20	0.14	0.23	0.16
Potato	1.19	1.53	1.70	1.90	1.40	1.66	3.74
Other crops	3.70	3.03	3.54	5.54	5.15	6.00	4.83

Note: * Data from the year 2003-04 onwards excludes Jharkhand while figures for all other years are for Bihar prior to reorganisation.

Source: Directorate of Statistics and Evaluation, Government of Bihar.

crops, other than food grains and cotton, was only 12 per cent. Thus, despite the recommendations of the Johl Committee (1986) in Punjab for diversification of crops, from rice-wheat rotation to other high value crops like fruits and vegetables and other commercial crops, as early as 1986, the actual cropping pattern in Punjab has not changed. Wheat and rice have maintained their share of area throughout the nineties (Table 2.4(a)). The situation was even worse in Bihar where apart from food crops, the share of other commercial crops, vegetables and fruits was only 13 per cent.

The reason for the dominance of rice and wheat in the total cropped area becomes clear when we look at the relative productivity of selected crops in Punjab and Bihar (Table 2.5). The productivity of these two crops was highest among all the food crops in both the states. In addition, the hike in productivity of these two superior cereals was also higher than that of coarse cereals as well as pulses. Price of these two main cereals was also higher than other coarse cereals, due to which the profitability remained always higher for them, thus leading to dominance of these crops in the cropping pattern. Comparing these two states, productivity of rice and wheat in Punjab was more than double that of Bihar. In the case of other coarse cereals, the difference was not substantial, while in the case of pulses, productivity in Bihar was at par with Punjab, or even higher in the case of *kharif* pulses, as was also the case with *kharif* oilseeds (see Table 2.5). The productivity of sugarcane was also 30 per cent higher in Punjab compared to Bihar, although its percentage in gross cropped area was only miniscule.

The reasons for such differences in productivity become clearer when we look at the level of irrigation in Punjab and Bihar. Around 98 per cent of gross cropped area in Punjab was irrigated, compared to 61 per cent in Bihar (Table 2.1). Thus, Punjab is almost fully irrigated and productivity of irrigated land for rice and wheat is much higher than non-irrigated land. Further, nitrogen fertiliser which is used for the growth of the plant as well as to enhance yield rate, was applied in large quantities on rice and wheat and its consumption per hectare was almost double in Punjab compared to Bihar (Table 2.7).

Table 2.5

Relative Productivity of Selected Agricultural Commodities

(Kgs per hectare)

	TE 1982-83	*TE 1987-88*	*TE 1992-93*	*TE 1997-98*	*TE 2000-01*	*TE 2008-09*	*TE 1982-83*	*TE 1987-88*	*TE 1992-93*	*TE 1997-98*	*TE 2000-01*	*TE 2007-08*
	Punjab						**Bihar**					
Rice	2947	3223	3293	3331	3335	3970	830	1060	987	1384	1384	1266
Wheat	2890	3347	3760	3989	4530	4393	1350	1567	1790	2170	2127	1861
CC (*kharif*)	1651	1666	1968	1950	2505	2967	877	1121	1215	1249	1408	1287
CC (*rabi*)	1746	2177	2761	3125	3384	-	719	757	1740	2492	2465	2976
Pulses (*kharif*)	700	857	764	751	629	783	603	673	701	705	884	873
Pulses (*rabi*)	513	704	722	834	878	924	604	700	775	717	842	741
Oilseeds (*kharif*)	898	770	570	569	506	488	395	427	436	599	678	938
Oilseeds (*rabi*)	648	977	1276	1312	1184	1398	486	588	690	700	798	1007
Cotton	307	483	547	367	317	715	-	-	-	-	-	-
Sugarcane	58000	60900	59900	62000	66400	59740	33100	35000	48900	44900	44000	41424
Jute & *mesta*	-	-	-	-	-	-	982	1285	1343	1555	1202	1725

Note: CC - coarse cereals.
Same as table 2.4(b).

Source: Agriculture Statistics at a Glance, Directorate of Economics & Statistics.

Table 2.6

Percentage of Irrigated Area under Principal Crops

	Bihar		*Punjab*		*All India*	
	(1996-97)	*(2007-08)*	*(1993-94)*	*(2007-08)*	*(1997-98)*	*(2007-08)*
Rice	40.4	57.0	95.0	99.4	50.2	56.9
Bajra	-	-	62.5	93.5	5.9	9.5
Maize	47.4	60.3	58.8	64.4	21.1	23.5
Wheat	89.0	91.9	94.8	98.6	85.0	90.9
Barley	16.3	-	94.6	97.8	57.3	-
Total cereals	52.9	68.7	93.8	98.1	47.3	-
Gram	3.3	12.4	53.8	82.4	21.2	34.1
Arhar (tur)	-	-	81.8	93.2	5.6	4.4
Total pulses	2.1	3.2	89.8	87.5	11.8	16.2
Total food grains	47.8	63.1	93.8	98.0	40.6	46.8
Groundnut	-	-	50.0	21.4	19.6	20.3
Rapeseed & mustard	37.6	39.8	87.4	86.8	57.6	75.5
Soyabean	-	-	-	-	2.7	1.2
Sunflower	-	83.2	68.7	100.0	21.4	31.8
Total oilseeds	20.2	38.2	62.2	87.2	24.4	27.1
Sugarcane	30.6	23.7	75.1	96.2	92.6	93.5
Cotton	-	-	99.6	99.2	36.3	35.1
Tobacco	66.7	78.1	-	-	41.3	52.6
All crops	46.6	60.6	91.7	97.7	38.2	44.6

Note: Same as Table 2.4(b).

Source: Directorate of Statistics and Evaluation, Government of Bihar.

Table 2.7

Consumption of Fertilisers in 2008-09

(Kgs per hectare)

State	*Nitrogen*	*Phosphorous*	*Potassium*	*Total*
Bihar	123.77	33.37	21.83	178.98
Punjab	166.83	47.51	7.08	221.42

Source: *Fertiliser Statistics, 2010-11.*

Table 2.8

Changes in Various Farm Energy Sources in Punjab and Bihar

(Numbers)

Year	*Implements*	*Bihar*	*Punjab*	*All India (Average)*
1971-72[a]				
	Working bovine[b]	90	47	68
	Tractor[c]	4	107	7
	Pump (electric/diesel)	1	8	3
1981-82				
	Working bovine[b]	94	34	62
	Tractor[c]	25	399	41
	Power tiller[c]	1	478	21
	Pump (electric/diesel)	5	23	7
1991-92				
	Working bovine[b]	115	20	59
	Power tiller[c]	59	1024	109
	Tractor[c]	25	584	41
	Pump (electric/diesel)	10	42	12

Note: a : 1971-72, there were no power tillers.
b : Working bovine per 100 hectares of operated area.
c : Per 10,000 hectares of operated area.

Source: *Agricultural Research Data Book 2004*, Indian Council of Agriculture Research.

The domination of medium and large landholders in operational holdings in Punjab, led by higher irrigation, use of power and mechanisation (Table 2.8), resulted in higher yield in Punjab, compared to most of the other states in India. The development of a regulated market system ensured a much larger proportion of production of rice and wheat arriving at the market place in Punjab. The market arrival of rice was as high as 83 per cent and that of wheat was 51 per cent in TE 1999-2000 (Table 2.9). In contrast, in the case of Bihar, the market arrival of rice and wheat was merely 25 and 17 per cent, respectively in TE 1999-2000. This should not be taken as the marketed surplus because a large proportion of production of rice and wheat in Bihar is sold in the villages itself and in unorganised markets.

Table 2.9

Market Arrival of Wheat and Rice in Selected States

(Per cent of production)

State	Rice			Wheat		
	TE 1982-83	TE 1991-92	TE 1999-2000	TE 1982-83	TE 1991-92	TE 1999-2000
Andhra Pradesh	42.03	42.33	70.97	-	-	-
Bihar	15.97	14.87	24.5	13.87	15.2	17.13
Gujarat	42.8	50.23	60.2	44.23	45.7	63.2
Haryana	92.3	70.7	77.33	41.4	41.93	44.57
Karnataka	18.43	23.77	44.13	8.33	8.93	29.07
Kerala	17	8.53	26.3	-	-	-
Madhya Pradesh	16.27	14	45.5	10.77	10.77	40.5
Maharashtra	15.47	22.1	45.83	29.7	30.03	59.03
Orissa	4.57	5.97	5	-	-	-
Punjab	90.37	84.8	82.83	48.43	47.27	51.33
Rajasthan	35.77	27.97	62.5	18.97	18.87	20.33
Tamil Nadu	35.33	34.47	43.57	-	-	-
Uttar Pradesh	26.23	28.5	40.13	17.2	18.6	23.4
West Bengal	16.57	15.97	15.53	-	-	-
All India	31.23	30.23	41.53	26.03	26.53	34.17

Source: *Bulletin on Food Statistics*, Ministry of Agriculture, various issues.

Table 2.10 presents the data showing the role of official agencies in procurement of rice and wheat from regulated markets in different states. In Bihar, the role of official agencies in procurement was almost negligible compared to the overwhelming role played by them in Punjab, procuring 85 per cent of rice and more than 90 per cent of wheat of all market arrivals, in the late nineties. The absence of any large scale procurement operation by the government agencies in Bihar has resulted in a relative decline in prices of rice and wheat in that state, compared to Punjab after the mid-nineties when minimum support prices (MSP) of paddy and wheat were substantially hiked. This shows that even with low yield levels the farmers in Bihar were not getting the benefits of the increased MSP.

Table 2.10

Market Arrival of Rice and Wheat Procured

(Per cent)

State	Rice		Wheat	
	TE 1991-92	TE 1999-2000	TE 1991-92	TE 1999-2000
Andhra Pradesh	65.0	65.7	–	–
Bihar	0.5	0.5	–	–
Haryana	77.4	59.3	82.1	80.1
Madhya Pradesh	56.2	31.1	–	11.4
Punjab	85.0	84.0	93.3	90.0
Rajasthan	78.0	8.3	12.0	38.8
Tamil Nadu	42.9	28.4	–	–
Uttar Pradesh	46.9	20.3	30.9	23.2
West Bengal	5.7	8.7	–	–
All India	50.9	41.5	61.5	49.1

Note: – Indicate negligible or nil procurement.
Source: *Bulletin on Food Statistics*, Ministry of Agriculture, various issues.

Table 2.11

Farm Harvest Prices of Major Crops in Bihar and Punjab

(Rs/quintal)

Year	Paddy			Wheat			Maize		
	Bihar	Punjab	Difference	Bihar	Punjab	Difference	Bihar	Punjab	Difference
1990-91	214	246	32	310	217	-93	236	278	42
1991-92	256	362	106	381	242	-139	302	315	13
1992-93	297	322	25	384	321	-63	321	360	39
1993-94	310	368	58	388	353	-35	296	387	91
1994-95	330	390	60	400	365	-35	335	429	94
1995-96	356	418	62	443	387	-56	363	488	125
1996-97	370	430	60	491	429	-62	399	414	15
1997-98	377	446	69	499	502	3	395	472	77
1998-99	380	482	102	503	550	47	398	547	149
2007-08	540	720	180	575	723	148	938	897	41

Source: Various issues of *Farm Harvest Prices in India*, Ministry of Agriculture, Directorate of Economics & Statistics.

Agrarian Structure[2]

As described earlier, the household data in this study is based on a sample survey conducted in 12 villages in Bihar and 16 villages in Punjab for a total number of 3,446 households in Punjab and 3,043 households in Bihar, from which around 800 households were selected as sample for detailed survey. In this section, we broadly look into the main characteristics of the Census data for these two states.

Table 2.12 presents a sketch of households covered in the Census across various size classes in the two states. As was pointed out earlier, a majority of the households in Bihar belonged to the marginal farmer households, while in Punjab the majority belonged to the landless households. Among the cultivating households, the number of marginal farmers was much higher in Bihar compared to Punjab, and number of large farmers was lower than that of Punjab.

Table 2.13 presents the percentage distribution of households across various districts in Punjab and Bihar. It is seen from the data that the Census households covered almost all the major regions of these two states. Table 2.14 presents the distribution of households across various castes. The distribution of landless households was skewed towards scheduled castes (SC) and other backward castes (OBCs) in both the states. Around 74 per cent of the landless households in Punjab belonged to the SC category, while the proportion of OBC and the upper castes (General) was 16 and 10 per cent, respectively. In the case of Bihar, 42 per cent of landless labourers belonged to the SC category, while 39 per cent belonged to OBC. The upper castes generally dominated the land ownership: In Punjab 86 per cent of large farmers belonged to the upper castes while number of large farmers among the SCs was less than 2 per cent. The picture was almost the same in Bihar as well. Among various religions, Sikhs dominated in Punjab with 78 per cent of the households while in Bihar the majority belonged to Hindus whose ratio was 94 per cent among the Census households. Hindus were the other dominant community in Punjab with 20 per cent households and Muslims were the second dominant community accounting for the remaining 6 per cent of the

2. This Section is based on the Census carried out by our team.

households in Bihar. Christians were almost non-existent in both the states (table 2.15).

Table 2.12

Distribution of Census Households across Various Farm Sizes

(Number of households)

	Marginal	*Small*	*Medium*	*Large*	*Landless*	*Total*
Punjab						
	619 (18.0)	325 (9.4)	255 (7.4)	182 (5.3)	2065 (59.9)	3446 (100.0)
Bihar						
	1566 (51.5)	264 (8.7)	119 (3.9)	54 (1.8)	1040 (34.2)	3043 (100.0)

Note: Figures in parentheses are respective percentages of total.
Source: Census by our research team.

Table 2.13

Distribution of Census Households by Districts

(Per cent)

	Marginal	*Small*	*Medium*	*Large*	*Landless*	*Total*
Punjab						
Amritsar	10.7	11.1	16.5	11.5	23.0	18.5
Bathinda	9.7	11.4	7.5	17.6	12.2	11.6
Hoshiarpur	19.9	10.5	8.6	3.8	10.7	11.8
Jalandhar	3.6	4.9	6.3	2.7	9.7	7.5
Mansa	12.4	17.2	20.0	18.1	8.9	11.6
Moga	9.4	16.0	15.3	20.9	13.6	13.6
Patiala	18.4	12.3	11.4	9.3	9.1	11.3
Sangrur	16.0	16.6	14.5	15.9	12.9	14.1
Bihar						
Bhojpur	20.6	33.0	23.5	25.9	23.9	23.0
Gopalganj	38.3	13.3	10.9	9.3	7.8	24.1
Jahanabad	16.6	23.1	25.2	14.8	32.1	22.8
Khagaria	24.5	30.7	40.3	50.0	36.2	30.1

Source: Census by our research team.

Table 2.14

Distribution of Census Households across Caste Groups

(Per cent)

	General	OBC	SC	ST	Others
Punjab					
Marginal	49.9	42.3	7.8	0.0	0.0
Small	70.2	24.9	4.9	0.0	0.0
Medium	74.1	22.4	3.5	0.0	0.0
Large	86.3	12.1	1.6	0.0	0.0
Landless	9.7	16.0	74.3	0.0	0.0
Total	31.5	21.8	46.7	0.0	0.0
	General	OBC-2	OBC-1	SC	ST
Bihar					
Marginal	16.7	55.7	11.4	8.1	8.2
Small	41.3	52.7	1.9	1.1	3.0
Medium	57.1	37.0	1.7	0.8	3.4
Large	35.2	59.3	0.0	0.0	5.6
Landless	3.2	38.8	12.3	41.7	4.0
Total	16.1	49.0	10.3	18.6	6.1

Note: OBC-2 indicates relatively more backward and OBC-1 are relatively less backward with most of them having ownership of land, e.g., Yadav, Kurmi etc.

Source: Census by our research team.

Table 2.15

Distribution of Census Households across Religions

(Per cent)

	Hindu	*Muslim*	*Sikh*	*Christian*
Punjab				
Marginal	22.8	1.9	75.3	0.0
Small	13.8	0.0	86.2	0.0
Medium	7.8	0.0	92.2	0.0
Large	3.8	1.1	95.1	0.0
Landless	22.6	3.5	73.5	0.5
All	19.7	2.5	77.5	0.3
Bihar				
Marginal	91.8	8.2	0.0	0.0
Small	97.0	3.0	0.0	0.0
Medium	96.6	3.4	0.0	0.0
Large	94.4	5.6	0.0	0.0
Landless	96.0	4.0	0.0	0.0
All	93.9	6.1	0.0	0.0

Source: Census by our research team.

Table 2.16 presents the pattern of tenancy among the Census households. Comparing Punjab and Bihar, it is seen from the statistics that in Punjab the distribution of leased-in land was almost even among all size classes, whereas in Bihar, 85 per cent of leased-in land was in favour of the marginal farmers, with the large farmers accounting for hardly any leased-in land. On the contrary, the distribution of leasing-out land was more equitable in Bihar as compared to Punjab. In aggregate, more land area was leased-in than leased-out in Punjab, while it was the converse in Bihar where more area was leased-out than leased-in.

Table 2.16

Pattern of Tenancy among Census Households

Landholding	*Per cent Leased-in to Total Owned Area*	*Per cent Leased-out to Total Owned Area*	*Percentage of Leasing in to Total Leasing-in Area*	*Percentage of Leasing-out to Total Leasing out Area*
Punjab				
Marginal	42.0	16.0	23.06	11.22
Small	25.4	18.3	19.83	18.26
Medium	24.7	10.7	31.11	17.26
Large	9.9	15.9	26.00	53.25
Total	19.2	15.0	100.00	100.00
Bihar				
Marginal	35.61	10.85	84.51	21.71
Small	5.45	18.23	9.97	28.14
Medium	0.58	19.21	1.06	29.73
Large	2.63	14.28	4.46	20.41
Total	12.93	15.33	100.00	100.00

Source: Census by our research team.

Table 2.17 presents the employment pattern in Punjab and Bihar. Among rural households, around 26 per cent in Punjab and 37 per cent in Bihar were cultivators operating some land area, either owned or tenanted. Another 20 per cent households in Punjab and 14 per cent in Bihar were employed as agricultural labourers. Thus, around 45 to 50 per cent households in both the states were directly employed in agriculture and related activities. Among non-agriculture activities, secondary and tertiary activities like construction, transportation etc., employed around 21 per cent households in Punjab and 22 per cent in Bihar. Public and private sector jobs accounted for around 11 per cent households in both the states.

Self-employment in business and crafts was next in importance providing employment to 13 per cent households in Punjab and 9 per cent in Bihar. Attached labour and other skilled labour in non-agricultural activities were other minor sources of employment to the rural households. Among the landless households, only 30 per cent were employed in agriculture and the remaining 70 per cent earned their bread and butter from non-agricultural activities—non-agricultural labour and self-employment in business being the chief non-agricultural activities for these households in both the states.

Table 2.17

Employment Pattern among the Census Households

(Per cent of households)

	Marginal	*Small*	*Medium*	*Large*	*Landless*	*Total*
Punjab						
Self-employed in agri.	35.70	71.38	86.67	89.56	2.03	25.51
Agriculture labour	15.02	1.54	0.00	0.00	28.04	19.65
Non-agri. labour	12.44	2.46	0.00	0.00	30.56	20.78
Attached labour	0.00	0.00	0.00	0.00	5.13	3.08
Pubic service	7.27	5.23	4.31	1.65	6.30	5.98
Private service	6.95	4.62	2.75	0.55	6.63	5.89
Caste occupation	0.32	0.00	0.00	0.00	1.84	1.16
Self-emp. in business	15.51	10.15	2.35	1.65	14.92	12.94
Skilled worker	2.10	0.31	0.00	0.00	1.65	1.39
Transfer earners	4.68	4.31	3.92	6.59	2.91	3.63
Bihar						
Self-employed in agri.	46.42	75.00	74.79	83.33	5.67	36.74
Agriculture labour	9.20	0.38	0.00	0.00	26.54	13.84
Non-agri. labour	13.41	0.00	0.00	0.00	44.81	22.21
Attached labour	0.06	0.00	0.00	0.00	0.00	0.03
Pubic service	2.11	3.03	3.36	3.70	0.58	1.74
Private service	10.98	17.05	14.29	11.11	5.58	9.79
Caste occupation	4.92	1.14	0.00	0.00	5.00	4.34
Self-emp. in business	9.45	1.52	5.04	1.85	10.29	8.74
Skilled worker	2.11	0.00	0.00	0.00	1.15	1.48
Transfer earners	1.34	1.89	2.52	0.00	0.38	1.08

Source: Census by our research team.

Among the assets owned by the households, the major productive ones were tractors and tractor related implements, combine harvesters, electric generators etc. The consumer assets, which are also indicators of social status of the households, included telephones, refrigerators and air-conditioners. In Punjab, the television was the most common asset followed by the refrigerator and the telephone owned by the largest number of households (Table 2.18). Among productive assets, the tractor was the most common, while a few households also owned a car or a jeep for their day-to-day use. However, Bihar lagged far behind Punjab in both productive as well as consumer assets. In comparison to two-thirds of the households in Punjab who owned televisions, only 9 per cent of the households owned televisions in Bihar. Similarly, only 3 per cent of the households in Bihar, in comparison to 14 per cent in Punjab, owned a tractor that is the most productive asset in modern day agriculture. On an average and not surprisingly, the lifestyle reflected by the consumer and productive assets in Punjab, was far better and wealthier than that in Bihar.

Table 2.18

Proportion of Households Owning Productive and Consumer Assets

(Per cent of households)

	Marginal	Small	Medium	Large	Landless	Total
Punjab						
Tractor	6.85	25.53	59.57	91.34	0.86	13.81
Crop harvester	0.20	0.35	0.85	3.90	0.00	0.38
Reaper	0.00	1.06	1.70	12.99	0.05	1.10
Thresher	0.40	3.19	8.09	23.38	0.27	2.61
Jeep/car	5.04	4.61	7.66	21.21	1.63	4.09
Truck/bus	0.81	0.00	0.00	0.87	0.05	0.20
T.V.	68.75	77.30	83.83	90.91	56.72	64.28
Refrigerator	38.51	51.42	62.98	81.39	23.89	34.76
Air-conditioner	0.00	0.71	0.43	3.90	0.18	0.46
Electric generator	1.41	2.13	7.66	22.51	0.86	2.96
Telephone	29.03	37.94	43.83	72.29	14.17	24.17
Bihar						
Tractor	1.27	7.52	18.92	38.64	0.20	2.73
Crop harvester	0.00	0.00	0.00	2.27	0.00	0.03
Reaper	0.06	0.65	0.00	2.27	0.00	0.13
Thresher	1.64	12.42	27.93	54.55	0.10	3.94
Jeep/car	0.38	1.96	4.50	2.27	0.20	0.66
Truck/bus	0.06	0.33	1.80	2.27	0.00	0.16
T.V.	7.15	20.59	35.14	59.09	3.10	8.94
Refrigerator	0.06	0.98	0.00	2.27	0.10	0.20
Air-conditioner	0.06	0.00	0.00	0.00	0.00	0.03
Electric generator	0.19	0.98	2.70	2.27	0.00	0.33
Telephone	1.39	6.54	12.61	40.91	0.90	2.73

Source: Census by our research team.

The living standard or economic well-being of the households is also reflected in the kind of dwelling houses. Table 2.19 gives a picture of the type of houses owned by the Census households. In Punjab, around 95 per cent of the households had *pucca* or semi-*pucca* houses, usually of bricks, and only 5 per cent of the households were staying in mud houses. In Bihar, 42 per cent of the households were staying in mud houses and another 19 per cent had mixed houses made of bricks and mud. Likewise, 88 per cent of the houses in Punjab had domestic electricity connections, whereas in Bihar, only 27 per cent of the households had domestic electricity connections. Thus, the majority of the houses in Bihar did not have electricity connections. This also explains why households in Bihar lacked consumer assets like televisions and refrigerators, which have become essential rather than luxury items in rural as well as urban areas in most parts of the country. Table 2.20 also reveals the poor agricultural infrastructure in Bihar. Whereas in Punjab, around 60 to 70 per cent of the cultivators had electric tube-well connections, in Bihar less than 10 per cent of the large farmers and less than 1 per cent of the medium, small and marginal farmers had electric tube-well connections.

Table 2.19

Quality of Housing across Various Farm Size Classes

(Per cent)

	Pucca	*Kachha*	*Mixed*	*Rented*
Punjab				
Marginal	48.19	1.41	50.40	0.00
Small	47.87	0.71	51.42	0.00
Medium	49.36	0.43	50.21	0.00
Large	71.86	0.00	28.14	0.00
Landless	32.61	8.04	58.81	0.54
Total	39.87	5.43	54.35	0.35
Bihar				
Marginal	40.67	39.60	19.73	0.00
Small	53.27	23.53	23.20	0.00
Medium	63.96	10.81	25.23	0.00
Large	79.55	6.82	13.64	0.00
Landless	29.07	55.84	15.08	0.00
Total	39.53	41.80	18.67	0.00

Source: Census by our research team.

Table 2.20

Proportion of Households by Types of Electrical Connections

(Per cent of households)

Operated Land	Domestic	Tube Well	Comm./ Business	Industrial	Any Other
Punjab					
Marginal	89.92	23.99	0.20	0.00	0.00
Small	94.33	47.16	0.00	0.00	0.00
Medium	94.04	56.17	0.43	0.00	0.43
Large	93.94	70.13	0.00	0.00	0.00
Landless	86.01	3.00	0.18	0.05	0.05
Total	88.33	17.76	0.17	0.03	0.06
Bihar					
Marginal	25.49	0.51	0.00	0.00	0.00
Small	39.87	1.31	0.00	0.00	0.00
Medium	51.35	0.00	0.00	0.00	0.00
Large	68.18	9.09	0.00	0.00	0.00
Landless	21.08	0.00	0.00	0.00	0.00
Total	27.05	0.53	0.00	0.00	0.00

Source: Census by our research team.

Finally, Table 2.21 and 2.22 present different activities through which households were diversifying their occupations and trying to increase their earning opportunities. Table 2.21 presents the share of households engaged in animal husbandry to supplement their farm income. Almost all categories of households in both the states were engaged in rearing buffaloes/cows for the milk. The percentage of households keeping milch animals had a direct relationship with farm size, as farmers with bigger farms had larger flock compared to small farmers in both the states. However, a majority of the households kept only one or two buffaloes/cows, probably to satisfy domestic needs. The households keeping more than six such animals, most likely for commercial purposes, were less than 5 per cent in Punjab and less than 1 per cent in Bihar.

Table 2.21

Ownership of Livestock

	% HH having Cows or Buffaloes	*Percentage Distribution of those having Cows or Buffaloes by Numbers*					
		1	*2*	*3*	*4-5*	*6-10*	*10+*
Punjab							
Marginal	59.07	37.88	34.47	15.70	8.87	2.73	0.34
Small	73.05	21.84	35.92	23.79	12.62	4.37	1.46
Medium	80.43	11.64	34.92	21.69	23.28	6.88	1.59
Large	85.28	10.15	15.74	16.24	38.07	16.24	3.55
Landless	35.60	65.43	27.17	4.08	2.17	0.89	0.26
Total	48.43	42.60	29.06	11.98	11.26	4.13	0.96
Bihar							
Marginal	46.05	70.47	22.12	5.63	1.51	0.27	0.00
Small	70.92	58.06	29.49	5.07	6.45	0.92	0.00
Medium	81.08	53.33	25.56	8.89	11.11	0.00	1.11
Large	81.82	41.67	25.00	5.56	22.22	5.56	0.00
Landless	19.28	83.42	12.95	3.11	0.52	0.00	0.00
Total	41.54	68.28	22.31	5.38	3.48	0.47	0.08

Source: Census by our research team.

Farmers try to diversify their income by changing cropping pattern, supplementing income through other agriculture & allied activities like dairy and poultry farming, fishing, bee-keeping and through other non-agriculture activities, like operation of retails shops in the village or nearby town, small industry, business or trading activities. Among our Census households in both the states, rearing milch animals was the only major activity undertaken by the farmers either for supplementing income through selling milk or rearing buffaloes and cows for their personal consumption of milk.

The most desired crop diversification in Punjab was hardly reflected among the Census households, as the percentage of those growing high value cash crops like fruits, vegetables and flowers was lower than 5 per cent, though 20 per cent of the households in the category of large farmers were growing high value crops. In Bihar, the proportion of such households was less than 1 per cent on an average, and less than 5 per cent in the case of large farmers, who had the real potential to diversify into such crops. Among agriculture & allied activities, less than 2 per cent of the households in Punjab and less than 1 per cent in Bihar were engaged in fishing, poultry farming and bee-keeping. Similar was the case of other agro-based small industries. Thus, only 6 per cent of the households in Punjab and 4 per cent in Bihar were able to diversify their income through supplementary activities other than the main crops and dairy farming, which remained the major source of employment as well as earning of the selected rural households. This apparently also reflected the rural picture in totality.

Table 2.22

Diversification of Activities

(Percentage of households)

	Marginal	*Small*	*Medium*	*Large*	*Landless*	*Total*
Punjab						
Growing vegetables	2.26	2.98	5.53	10.83	0.00	1.66
Growing flowers	1.69	2.32	3.56	9.58	0.00	1.31
Growing fruits	0.19	0.00	0.00	0.42	0.00	0.05
Milk vendors	0.56	1.66	0.00	1.67	0.30	0.52
Fishing	0.19	0.00	0.00	0.00	0.17	0.14
Poultry	1.88	2.98	3.56	10.00	0.21	1.55
Bee-keeping	0.19	0.00	0.00	0.00	0.00	0.03
Shop keepers	0.00	0.33	0.00	0.00	0.09	0.08
Oilseed crushing	1.32	0.00	0.00	0.00	1.11	0.90
Other agro-based	0.19	0.00	0.00	0.00	0.09	0.08
Total other than milch	8.47	10.26	12.65	32.50	1.96	6.31
Keeping milch animals	59.07	73.05	80.43	85.28	35.6	48.43
Gross total	67.54	83.31	93.08	117.78	37.56	54.74
Bihar						
Growing vegetables	1.20	1.31	3.60	2.27	0.10	0.95
Growing flowers	0.00	0.00	0.90	0.00	0.00	0.03
Growing fruits	0.00	0.65	0.90	2.27	0.10	0.16
Milk vendors	0.00	0.00	0.90	0.00	0.00	0.03
Fishing	0.00	0.00	0.00	0.00	0.00	0.00
Poultry	0.00	0.33	0.90	0.00	0.10	0.10
Bee-keeping	0.06	0.33	0.00	2.27	0.00	0.10
Shopkeepers	2.59	2.29	7.21	0.00	1.00	2.17
Oilseed crushing	0.25	0.65	0.90	0.00	0.00	0.23
Other agro-based	0.76	0.33	0.00	0.00	0.00	0.43
Total other than milch	4.87	5.88	15.32	6.82	1.30	4.21
Keeping milch animals	46.05	70.92	81.08	81.82	19.28	41.54
Gross total	50.92	76.8	96.4	88.64	20.58	45.75

Source: Census by our research team.

We also came across a few cases of contract farming in Punjab. These households belonged mostly to the large farmer categories. Though one or two cases of medium and small farmers were also covered. Although in Punjab, contract farming is being practiced for many crops like potato, *basmati*, maize, *hyola*, a few pulses and oilseeds, in seeds and medicinal plants, we could study the cases of potato, *basmati* and maize. The agri-business firms with which farmers were having contracts were Pepsi, Hindustan Lever and the Punjab Agro Foodgrain Corporation Ltd. (PAFC) in the case of maize. As our data was limited to socioeconomic and employment aspects only, it was not possible to get a glimpse of the economics of production and resource-use efficiency of these contracting households. Another limitation of our Census data was that in both the states we did not find any farmer growing any crop for export purposes or selling his/her produce overseas through any intermediaries.

Annexure Table A-2.1

Area under Fruits, Vegetables and Floriculture in the Selected States

(Hectares)

	Fruits		*Vegetables*		*Flowers*	
	Bihar	*Punjab*	*Bihar*	*Punjab*	*Bihar*	*Punjab*
1991-92	26692	7266	84331	845	-	-
1992-93	27117	7652	88731	8529	85	332
1993-94	28284	8177	91397	1025	-	-
1994-95	27899	8177	80773	10250		
1995-96	28565	8442	85685	10500	104	550
1996-97	292789	90295	602439	112935	104	550
1997-98	299799	90295	603470	120135	86	560

Source: *Agricultural Research Data Book*, 2001.

3 Production Structure and Resource Use

The previous chapter presented a brief comparative picture and the broad contours of ongoing changes in the agriculture sector in Punjab and Bihar. Based on state level secondary information and on our field Census data, the analysis indicated the broad changes that have occurred in the past decade in the crop and allied sectors in these two states, and their implications on the production structure, income and employment. Nonetheless, the secondary data fails to reveal the undercurrents and changes that are taking place at the grassroot level. These changes are better tracked by the micro level household survey data. Therefore, to get an insight of these changes and the socioeconomic implications of globalisation policies occurring at the national and global level on the village economy, a field survey was carried out with village and households as the primary units of investigation. In this chapter we present in brief the findings of our survey on cropping pattern, productivity and farm income structure in the two states under the purview of this study

Characteristics of the Region

Operational Land

Table 3.1 presents the statistics of the selected sample and their operated area. The pattern of land distribution observed for the selected cultivators presents the broad trends prevailing in these two states, which are also in consonance with the trends presented in the previous chapter. The number of households selected from each of the two states was almost the same, while distribution among different holdings varied, representing the variations in their numbers in the total population. As the percentage of marginal and small farmers was much higher in Bihar compared to Punjab, their representation in the sample was also much larger in Bihar than in Punjab. Conversely, the number of large holdings selected in Punjab was much higher compared to Bihar, representing the trend in the entire state. The average size of holdings among the selected households was 6.7 acres in Punjab that was much higher compared to 3.4 acres in Bihar.

Table 3.1

Distribution of Holdings and Operated Area among the Selected Farmers based on Operated Area

Category	*No of Households*	*%*	*House-hold Size*	*Percentage of Literate Persons in the Family*	*Net Operated Area (Acres)*	*%*
Punjab						
Marginal	77	18.1	5.21	66.48	1.60	4.3
Small	82	19.2	6.55	79.47	3.96	11.5
Medium	58	13.6	6.47	73.13	7.71	15.8
Large	85	20.0	8.38	77.40	22.83	68.4
Landless	124	29.1	5.39	56.58	0.00	0.0
Total	426	100.0	6.32	70.61	6.66	100.0
Bihar						
Marginal	175	40.7	7.25	63.78	1.17	14.2
Small	88	20.5	9.14	72.40	3.68	22.4
Medium	62	14.4	11.02	81.70	7.27	31.2
Large	29	6.7	11.97	88.85	16.05	32.2
Landless	76	17.7	5.89	57.79	0.00	0.0
Total	430	100.0	8.26	71.19	3.36	100.0

Source: Field Survey.

The number of marginal holdings was more than 40 per cent and occupied around 14 per cent of the operational area among the selected farmers in Bihar, compared to 14 per cent of holdings and 4 per cent area under these holdings among the sample households in Punjab. More than 75 per cent of the holdings and 68 per cent of the area were less than 10 acres in the former case while this ratio was 50 and 32 per cent, respectively in the latter case (Table 3.1). The astonishing fact that the number of marginal holdings is falling and of large holdings is increasing in Punjab, against the trends of almost all other states, is also proven by our sample. Whereas in Bihar the large holdings were less than 7 per cent, occupying only 32 per cent of area among the selected households, the number of large holdings in the Punjab sample was 20 per cent with 68 per cent operated area. Table 3.2 (third column) presents the average household land under tenancy by different size of cultivators. It is evident from the data that all households barring the category of landless in Punjab leased in more land then they leased out, thereby increasing the size of their operational holdings. Comparing different household categories, large farmers leased-in much more area, more than 5 acres, on an average, while small and marginal farmers leased in less than half acre. On the contrary, many of the small farmers were leasing out land in favour of medium and large farmers, a phenomenon in the literature called 'reverse tenancy', in Punjab as was also mentioned elsewhere. In Bihar on the other hand, it was the marginal and small farmers who leased in proportionately higher area compared to other categories, while large farmers were net losers as they were leasing out more land than they were leasing in. Because of the trends in 'reverse tenancy' in Punjab, the number of landless households selected was higher in Punjab compared to Bihar. In the case of cropping intensity in the two states, despite all ills, the selected cultivators in Bihar especially the marginal farmers, were trying hard to grow multiple crops so their cropping intensity compared very well with the selected households in Punjab.

Table 3.2

Characteristics of Operational Holdings

(Acres per household)

Category	*Own Land*	*Non-Cultivable Land*	*Net Leased-in*	*Encroached Area*	*Net Operated Area*	*Gross Cropped Area*	*Cropping Intensity*
Punjab							
Marginal	1.74	0.17	0.03	0.00	1.60	3.53	2.21
Small	3.13	0.24	1.03	0.04	3.96	8.72	2.20
Medium	5.82	0.13	2.01	0.00	7.71	16.17	2.10
Large	17.99	0.36	5.14	0.06	22.83	45.98	2.01
Landless	0.43	0.12	-0.31	0.00	0.00	0.00	0.00
Total	5.43	0.20	1.41	0.02	6.66	13.69	2.06
Bihar							
Marginal	1.05	0.04	0.16	0.01	1.17	2.81	2.39
Small	3.16	0.02	0.53	0.00	3.68	7.24	1.97
Medium	7.10	0.12	0.29	0.00	7.27	12.55	1.73
Large	16.38	0.21	-0.16	0.03	16.05	28.76	1.79
Landless	0.94	0.00	-0.94	0.00	0.00	0.00	0.00
Total	3.37	0.05	0.04	0.01	3.36	6.37	1.90

Source: Field Survey.

Household Size and Occupation Pattern

The household size was the highest in the case of large holdings with an average size of 8 members in Punjab and 12 members in Bihar. Marginal farmers and landless labourers had the lowest household size with 5 members each in Punjab, and 7 and 6 members, respectively in Bihar (Table 3.1). The direct relationship between household size and size of holdings was because of the prevalence of combined families among the large cultivators, and mostly nuclear families in the case of small and marginal farmers and landless labourers. The average literacy rate in both the states was almost similar (Table 3.1). On an average, around 71 per cent of the family members in the age group above six years in the family were literate in both the states.

The distribution of sample family members in different occupations following the 'National Classification of Occupation (1968)' is given in the proceeding tables. It is seen that the percentage of infants in the age group of 0-4 years among the total family members was around 7 per cent in the selected households in Punjab (Table 3.3), and around 13 per cent in Bihar (Table 3.4). The percentage of full time students either in school, college or doing other professional courses among the family members was almost the same in both the states, around 27 per cent. Thus, 35 to 40 per cent of the population in the selected sample in both the states was not in the working age. In addition, around 25 per cent of the family members in both the states were housewives or others engaged in household activities, and around 5 per cent were either unemployed or disabled/retired persons not having any active participation in any economic activity. Thus, around 35 to 40 per cent of the population was the truly active population among the sample households, and were engaged in some earning activity either in agriculture and related activities or in the secondary and tertiary sectors.

The trends in occupation depicted that out of 35 per cent of the working members in both the states, around 20 to 25 per cent were engaged in agriculture as cultivators or labourers and other allied activities like plantation, dairy etc. The remaining 10 to 15 per cent were engaged in the manufacturing/service sectors as government employees (on regular or ad hoc basis), transport operators or as construction workers. Due to the seasonal nature of employment and earnings in agriculture, most of the

Table 3.3

Occupational Distribution: Punjab

(Per cent of households)

	Marginal	*Small*	*Medium*	*Large*	*Landless*	*Total*
Main Occupation						
Cultivators	18.95	22.72	30.67	26.40	0.00	18.65
Agricultural labourers	4.49	0.37	0.00	0.00	24.21	6.70
Plantation, dairy supervisors	1.25	1.86	1.33	1.83	2.27	1.79
Unemployed	3.24	1.49	1.33	1.83	2.42	2.05
Students	22.19	27.93	24.53	29.35	31.16	27.77
Domestic duty	10.97	10.99	12.00	14.89	7.41	11.28
Other domestic services	15.96	15.27	13.07	11.52	13.16	13.55
Disabled persons	5.99	5.21	4.80	6.32	5.45	5.62
Infants (0-4)	7.98	7.26	8.27	4.21	8.62	7.04
Others	8.98	6.89	4.00	3.65	5.30	5.55
First Subsidiary Occupation						
Plantation, dairy supervisors	29.43	32.03	37.33	33.01	16.49	28.82
Cultivators	5.49	2.42	0.80	2.39	0.00	2.05
Agricultural labourers	1.50	1.68	0.00	0.00	2.87	1.27
Construction workers	0.50	0.19	0.00	0.00	3.03	0.86
Domestic duty	0.00	1.12	1.07	0.56	1.06	0.78
Retired/pensioners etc.	2.00	0.93	0.80	0.28	0.61	0.82
Others	2.00	0.74	1.87	1.12	3.03	1.75
All	40.90	39.11	41.87	37.36	27.08	36.34
Second Subsidiary Occupation						
Plantation, dairy supervisors	4.24	1.12	0.00	0.70	1.21	1.34
Cultivators	0.50	0.19	0.53	0.00	0.00	0.19
Agricultural labourers	0.25	0.56	0.00	0.00	0.91	0.37
Construction workers	0.00	0.00	0.00	0.00	0.76	0.19
Domestic duty	0.50	0.00	0.00	0.00	0.15	0.11
Other domestic services	0.00	0.00	0.27	0.00	0.15	0.07
Retired/pensioners etc.	0.25	0.56	0.53	0.14	0.00	0.26
Disabled persons	0.00	0.00	0.00	0.14	0.00	0.04
Others	0.25	0.56	0.00	0.14	0.30	0.26
All	5.99	2.98	1.33	1.12	3.48	2.83

Source: Field Survey.

Table 3.4

Occupational Distribution: Bihar

(Per cent of households)

	Marginal	*Small*	*Medium*	*Large*	*Landless*	*Total*
Main Occupation						
Cultivators	15.06	17.54	15.81	16.43	0.00	14.06
Agricultural labourers	3.31	0.00	0.00	0.00	11.09	2.55
Plantation, dairy supervisors	1.34	0.87	1.61	0.29	1.15	1.16
Construction workers	3.31	1.49	0.73	0.58	5.77	2.43
Transport operators	1.81	1.24	0.73	1.44	2.31	1.50
Students	20.58	27.24	33.09	36.60	24.48	26.56
Domestic duty	12.85	15.05	18.30	17.58	11.78	14.74
Other domestic services	13.17	11.07	6.73	7.20	11.09	10.61
Disabled persons	3.15	2.86	1.90	1.44	4.62	2.86
Infants (0-4)	15.30	12.81	9.96	8.65	16.17	13.15
Others	10.09	9.83	11.13	9.80	11.55	10.38
First Subsidiary Occupation						
Plantation, dairy supervisors	15.14	19.03	16.11	17.87	4.62	15.19
Cultivators	15.06	13.18	7.76	5.19	0.00	10.41
Farmers other than cultivators	0.16	0.00	0.44	0.58	3.00	0.57
Agricultural labourers	4.18	0.37	0.44	0.00	11.09	3.03
Construction workers	1.03	0.25	0.00	0.58	2.08	0.74
Transport operators	0.39	0.25	0.29	0.86	0.23	0.37
Domestic duty	0.00	0.00	0.73	1.73	2.54	0.62
Other domestic services	1.81	1.62	0.29	0.00	4.85	1.67
Others	1.34	1.37	2.49	1.15	2.08	1.64
All	39.12	36.07	28.55	27.95	30.48	34.23
Second Subsidiary Occupation						
Plantation, dairy supervisors	8.83	7.71	4.98	6.05	1.85	6.70
Cultivators	3.31	1.99	1.17	0.29	0.00	1.90
Agricultural labourers	1.18	0.37	0.00	0.00	1.15	0.65
Construction workers	0.47	0.00	0.00	0.00	0.46	0.23
Transport operators	0.24	0.00	0.15	0.29	0.00	0.14
Domestic duty	0.08	0.25	0.29	0.00	0.00	0.14
Other domestic services	0.47	0.25	0.15	0.00	0.00	0.25
Others	0.71	0.37	0.15	1.15	0.46	0.54
All	15.30	10.95	6.88	7.78	3.93	10.55

Source: Field Survey.

households had subsidiary occupation activities to support their families in the off-season. It is evident from the data that the percentage of family members engaged in the primary and secondary occupations almost tallied with the percentage of working population as mentioned above. Among the subsidiary activities, livestock and dairy farming were the leading ones. Almost all the cultivators and agricultural labourers and those others engaged in government employment, construction etc., kept milch animals for household milk consumption and in some cases for commercial purposes also. The farming occupation was the primary activity for cultivators in Punjab while in Bihar a significant number of cultivators had farming as their secondary occupation, possibly because of the low returns from agriculture in the state, as will be seen in the subsequent sections. Landless households and marginal farmers mostly played the role of agricultural labourers either as their main occupation, or as subsidiary activities in both the states.

Nature of Tenancy

There were three types of tenancy contracts prevailing in these two states under which land was given on lease, namely share cropping, fixed rent in cash and fixed rent in kind. The statistics presented in Table 3.5 reveals that Punjab's economy was highly monetised, where around 78 per cent of tenanted holdings were under fixed rent in cash contract and only 22 per cent of the contracts were in terms of kind or share cropping. In Bihar, on the other hand, a very high proportion of tenancy was still under barter terms, either in kind or in share cropping, and only 13 per cent of tenanted land was under fixed rent in cash. The average rent per acre was also much higher in Punjab (Rs 9,348) compared to Bihar (Rs 3,147).

Credit Market

The available literature on rural development indicates not only the shortage of cash flow to the farmers but also its irregularity. There are also reports that suggest that the funds are usurped usually by the well to do households, a majority of whom belong to large farmers (Kumar, 1991). Table 3.6 presents the amount of credit availed by the selected households during the pre- and post-reform periods. It is worth mentioning here that borrowings reported in the pre-reform period consist of the amount

Table 3.5

Tenancy Contracts in Punjab and Bihar

Category	No of Acres per Household			Amount Paid for Leasing-in (Rs per Acre)	Percentage of Holdings under		
	Share Cropping	Fixed Rent in Cash	Fixed Rent in Kind		Share Cropping	Fixed Rent in Cash	Fixed Rent in Kind
Punjab							
Marginal	0.01	0.19	0.00	7278	5.4	94.6	0.0
Small	0.15	0.90	0.06	8727	13.8	81.1	5.1
Medium	0.48	1.71	0.00	9118	22.0	78.0	0.0
Large	1.09	3.83	0.12	9623	21.7	75.9	2.3
Total	0.44	1.70	0.05	9348	20.3	77.5	2.2
Bihar							
Marginal	0.24	0.05	0.06	3281	68.4	14.4	17.2
Small	0.43	0.08	0.05	3318	77.2	13.4	9.4
Medium	0.26	0.06	0.09	2799	62.8	15.7	21.5
Large	0.45	0.00	0.17	2703	72.2	0.0	27.8
Total	0.31	0.06	0.07	3147	70.7	12.6	16.7

Source: Field Survey.

borrowed by the farmers before 1995; and borrowings in the post-reforms period are those that are reported thereafter. The borrowing during these two phases is not very comparable as data given are based on the interviewee's memory and therefore the pre-reform period would be incorrectly reported with the memory lag being more than 10 years for such data.

It is evident from the table that the per household credit needs of the large farmers were higher than that of the small farmers, and the amount of loan availed by the Punjab farmers was 10 times more than that availed by the Bihar farmers. On the other hand, the amount borrowed per acre had an inverse relationship with the farm size, whereby loans taken by marginal farmers were much higher compared to large farmers in both the states. In Punjab, the loan amount per household by marginal farmers was calculated as Rs 39,000, compared to Rs 211,000 by the large farmers. In Bihar, the amount borrowed by marginal farmers was less than Rs 6,000 while large farmers borrowed Rs 21,000 per household in the post-reform period. The loan amount varied from Rs 24,000 per acre to Rs 9,000 per

Table 3.6

Borrowings by Sample Households

Category	Marginal		Small		Medium		Large		Total	
	Before 1995	After 1995	Before 1995	After 1995	Before 1995	After 1995	Before 1995	After 1995	Before 1995	After 1995
Punjab										
					Rs per Household					
Institutional	5195	12558	3963	25335	3707	57276	5706	110741	3366	38351
Commission agent	1039	6455	4659	18829	0	34431	6059	57353	2293	21831
Traders/ML/landlord	1883	4883	0	8415	3448	9603	0	8200	927	7193
Friends/relatives	0	4221	0	5122	0	172	0	3565	153	3002
Own caste	0	1299	0	1646	0	5000	0	1941	23	1631
Upper caste	1558	7805	0	134	0	2069	0	5529	399	4466
Lower caste	0	0	0	610	0	0	0	0	0	124
Government programme	390	1299	0	9171	0	11914	0	23412	70	8446
Others	0	597	0	671	0	103	0	176	0	380
Total	10065	39117	8622	69933	7155	120569	11765	210918	7232	85424
					Rs per Acre					
Institutional	3245	7845	1001	6396	481	7433	250	4851	506	5762
Government programme	243	811	0	2315	0	1546	0	1026	11	1269
Commission agents	649	4032	1176	4753	0	4468	265	2513	345	3280
Traders/ML/landlord	1176	3050	0	2124	447	1246	0	359	139	1081
Friends/relatives	0	2637	0	1293	0	22	0	156	23	451
Own caste	0	811	0	416	0	649	0	85	4	245
Upper caste	974	4876	0	34	0	268	0	242	60	671
Lower caste	0	0	0	154	0	0	0	0	0	19
Others	0	373	0	169	0	13	0	8	0	57
Total	6287	24436	2177	17654	929	15647	515	9240	1087	12835

contd...

...contd...

Category	Marginal		Small		Medium		Large		Total	
	Before 1995	After 1995	Before 1995	After 1995	Before 1995	After 1995	Before 1995	After 1995	Before 1995	After 1995
Bihar										
					Rs per Household					
Institutional	114	451	1182	1545	0	4119	0	19069	372	2471
Traders/ML/landlord	74	2110	57	216	0	2016	0	0	109	1554
Friends/relatives	0	800	568	750	0	968	0	0	147	649
Own caste	0	634	0	625	0	1210	0	0	0	581
Upper caste	0	206	0	114	0	0	0	0	0	128
Lower caste	137	303	0	91	0	0	0	0	71	183
Government programme	451	499	420	68	0	1008	0	1372	342	1234
Others	0	571	1023	2273	0	0	0	172	265	709
Total	777	5575	3250	5682	0	9321	0	20614	1307	7508
					Rs per Acre					
Institutional	97	385	321	420	0	566	0	1188	238	735
Government programme	385	425	114	19	0	139	0	86	219	367
Traders/ML/Landlord	63	1798	15	59	0	277	0	0	70	462
Friends/relatives	0	681	154	204	0	133	0	0	94	193
Own caste	0	540	0	170	0	166	0	0	0	173
Upper caste	0	175	0	31	0	0	0	0	0	38
Lower caste	117	258	0	25	0	0	0	0	45	54
Others	0	487	278	617	0	0	0	11	170	211
Total	662	4749	883	1543	0	1282	0	1284	837	2233

Note: ML - Moneylender.

Source: Field Survey

acre from marginal farmers to large farmers in Punjab while it was recorded as Rs 4,700 and Rs 1,300 per acre, respectively in the case of marginal and large farmers in Bihar. Another notable point evident from the table is that institutional loans were also accessible to marginal and small farmers, although its absolute amount was much lower compared to the large and medium farmers.

Table 3.7 presents the percentage of amount borrowed by the households from institutional and non-institutional sources. It is evident from the data that in Punjab, marginal and small farmers had very good access to institutional loans and they competed very well with their counterpart large farmers, but the situation in Bihar was not as pleasant. In Punjab, 35 to 40 per cent of the loan requirements of the marginal and small farmers were fulfilled by institutional loans and other government lending programmes, compared to around 60 per cent in the case of large farmers. In Bihar on the other hand, institutional and other government borrowings met less than 20 to 25 per cent of the loan requirements of marginal and small farmers compared to above 95 per cent in the case of large farmers. These statistics point to the conclusion that whereas in Punjab, marginal and small farmers had good access to the institutional loans, in Bihar these loans were going largely to the medium and large farmers. In the case of non-institutional loans, commission agents (*kachha* and *pucca arhatias*) were the principal loan providers in Punjab, while trader/moneylenders were the main source of credit among the non-institutional sources.

Table 3.8 presents the statistics on the use of credit for productive and non-productive activities by the selected households. Of total borrowings, around 67 per cent in Punjab and 44 per cent in Bihar were used for agricultural purpose for buying seeds, fertilisers, and other nutrients, and for long-term purpose of installing tube-wells or buying tractors etc. Another 2 per cent of the borrowings in Punjab and 8 per cent in Bihar were used for productive purposes in non-agricultural activities. Among the non-productive activities, 25 per cent credit in Punjab and 43 per cent in Bihar was utilised in buying consumer durables, non-durables and for social ceremonies and the remaining 7 per cent in Punjab and 5 per cent in Bihar was used for litigation. Thus, around 70 per cent of the total credit in Punjab and 50 per cent in Bihar were used for productive purposes, which

Table 3.7

Details of Source of Household Credit

(Percentage of amount)

Category	Marginal		Small		Medium		Large		Total	
	Before 1995	After 1995	Before 1995	After 1995	Before 1995	After 1995	Before 1995	After 1995	Before 1995	After 1995
Punjab										
Institutional	51.6	32.1	46.0	36.2	51.8	47.5	48.5	52.5	46.5	44.9
Government programme	3.9	3.3	0.0	13.1	0.0	9.9	0.0	11.1	1.0	9.9
Commission agent	10.3	16.5	54.0	26.9	0.0	28.6	51.5	27.2	31.7	25.6
Traders/ML/landlord	18.7	12.5	0.0	12.0	48.2	8.0	0.0	3.9	12.8	8.4
Friends/relatives	0.0	10.8	0.0	7.3	0.0	0.1	0.0	1.7	2.1	3.5
Own caste	0.0	3.3	0.0	2.4	0.0	4.1	0.0	0.9	0.3	1.9
Upper caste	15.5	20.0	0.0	0.2	0.0	1.7	0.0	2.6	5.5	5.2
Lower caste	0.0	0.0	0.0	0.9	0.0	0.0	0.0	0.0	0.0	0.1
Others	0.0	1.5	0.0	1.0	0.0	0.1	0.0	0.1	0.0	0.4
Bihar										
Institutions	14.7	8.1	36.4	27.2	0.0	44.2	0.0	92.5	28.5	32.9
Government programme	58.1	8.9	12.9	1.2	0.0	10.8	0.0	6.7	26.2	16.4
Traders/ML/Landlord	9.6	37.9	1.7	3.8	0.0	21.6	0.0	0.0	8.3	20.7
Friends/relatives	0.0	14.4	17.5	13.2	0.0	10.4	0.0	0.0	11.3	8.6
Own caste	0.0	11.4	0.0	11.0	0.0	13.0	0.0	0.0	0.0	7.7
Upper caste	0.0	3.7	0.0	2.0	0.0	0.0	0.0	0.0	0.0	1.7
Lower caste	17.6	5.4	0.0	1.6	0.0	0.0	0.0	0.0	5.4	2.4
Others	0.0	10.3	31.5	40.0	0.0	0.0	0.0	0.8	20.3	9.4

Note: ML - Moneylender.
Source: Field Survey.

Table 3.8

End Use of Credit by the Sample Households

(*Percentage of amount*)

Category	*Marginal*		*Small*		*Medium*		*Large*		*Total*	
	Before 1995	*After 1995*	*Before 1995*	*After 1995*	*Before 1995*	*After 1995*	*Before 1995*	*After 1995*	*Before 1995*	*After 1995*
Punjab										
Agriculture	43.9	36.4	24.3	67.5	51.8	68.0	54.5	77.7	41.6	66.6
Non-agriculture	0.0	0.0	0.0	0.3	0.0	3.0	4.0	1.4	1.6	1.5
Non-durable	2.6	22.5	36.8	11.1	0.0	5.5	1.5	2.7	12.8	8.6
Durable	9.7	14.0	21.2	6.8	0.0	11.6	0.0	3.2	8.1	7.0
Social ceremony	25.8	21.2	17.7	3.3	48.2	5.4	40.0	8.9	31.3	9.5
Litigation	18.1	5.8	0.0	11.0	0.0	6.5	0.0	6.0	4.5	6.9
Bihar										
Agriculture	17.6	17.7	12.9	3.4	0.0	55.0	0.0	95.0	14.3	43.5
Non-agriculture	29.4	1.1	35.0	25.0	0.0	10.4	0.0	5.0	31.6	8.1
Non-durable	9.6	35.1	0.0	2.4	0.0	2.6	0.0	0.0	4.5	12.5
Durable	0.0	9.3	31.5	40.0	0.0	13.8	0.0	0.0	20.3	12.0
Social ceremony	43.4	27.5	19.2	29.2	0.0	16.4	0.0	0.0	28.4	18.5
Litigation	0.0	9.2	1.4	0.0	0.0	1.7	0.0	0.0	0.9	5.4

Source: Field Survey.

can be considered an investment that has larger significance in the development of the states in the long run. The rest of the credit was used for consumption and other purposes like social ceremonies etc.

Irrigation

An important reason for the Green Revolution spreading more successfully in Punjab and Haryana than in other states was the availability of assured irrigation in these two states. It was shown in the previous chapter that the net area irrigated in Punjab was around 94 per cent while the figure for Bihar was only 49 per cent. Among the households surveyed, the net irrigated area was around 99 per cent in Punjab while it was surprisingly very high—93 per cent—in Bihar.[1] However, comparing the source of irrigation in these two states, a very contrasting comparison came

Table 3.9

Area Irrigated in the Two States

(Per cent of NOA)

	Only Canal	*Canal and Tube Well (Electric/Diesel)*	*Only Electric Tube Well*	*Only Diesel Tube Well*	*Total*
Punjab					
Marginal	1.83	12.17	57.39	26.51	97.90
Small	2.16	8.77	64.32	23.42	98.66
Medium	1.12	14.21	73.20	10.61	99.14
Large	1.08	16.15	69.11	13.66	100.00
Total	1.24	14.57	68.70	14.86	99.36
Bihar					
Marginal	1.3	3.4	0.5	88.3	93.5
Small	1.7	1.9	0.1	90.4	94.0
Medium	5.4	3.3	0.0	83.1	91.8
Large	0.0	0.0	0.1	91.3	91.4
Total	2.7	2.4	0.1	88.2	93.5

Note: NOA - Net operated area.
Source: Field Survey.

1. The irrigated area reported by our investigators may not be irrigated area reported by official statistics. As the field area falls in high rainfall area lot of rain water accumulates in low land areas. This water is lifted using diesel-pump set with plastic pipeline. This officially may not be treated as irrigated area lout our survey included it as irrigated area.

to light. In Punjab, electric tube-wells were the main source of irrigation, which accounted for around 70 per cent of the total operated area. In Bihar, electric tube-wells were a minor source, visibly due to non-availability of power, while 88 per cent of the operated area was irrigated by diesel pump-sets. In Punjab, the contribution of diesel pump-sets was only 15 per cent. Diesel pump-sets were the obvious choice in Bihar as many of the rural areas were still to be electrified in the state. On the other hand, Punjab achieved total electrification long back and many of the state governments in the past provided free electricity for the agricultural purposes, giving a boost to electric pump-sets in the state. The contribution of canals in the irrigated area among the selected households in both the states was very low, only 2 per cent of the net operated area.

Crop-wise, paddy being a very high water intensive crop was fully irrigated in Punjab, while in Bihar it had 85 per cent area irrigated (Table 3.10).[2] The area under wheat was also mostly irrigated in both the states, as also were oilseeds, fruits and vegetables. Some coarse grains and pulses had significant un-irrigated area. In many of the cases around one-fourth of the area under the latter crops was un-irrigated in both the states. Probably that was also the reason for their low productivity as could be seen in the subsequent analysis.

High Yielding Variety Seeds

Like irrigation, high yielding variety (HYV) seeds constitute a major component of new technology. Punjab has remained the leading state in the use of HYV seeds. This is also reflected from the data presented in Table 3.11. Among the selected households, more than 90 per cent of the cropped area was under HYV seed in Punjab. Bihar lagged far behind in the use of HYV seeds as it had only 62 per cent of area under HYV. In addition to major food crops namely, wheat and paddy, cash crops like cotton, sugarcane, potato and oilseeds were almost fully covered by HYV seeds in Punjab. A few coarse grains, pulses and fodder crops were far behind in using HYV seeds in the state, probably because of lack of availability of

2. Whereas secondary sources showed that rice in Bihar was less than 50 per cent irrigated, our sample area under rice revealed 85 per cent irrigated, which may be explained by over reporting, confusing the irrigated area with that of rain-fed area as explained in the previous footnote and some sampling errors.

HYV seeds in these crops. In Bihar, among the major crops namely wheat, paddy, maize and gram, only wheat was fully covered by HYV seeds. Around three-fourth area of other major crops was under HYV seeds.

Table 3.10

Proportion of Area under Irrigation

	Marginal	*Small*	*Medium*	*Large*	*Total*
Punjab					
Paddy	100.0	99.0	100.0	100.0	99.9
Wheat	86.6	86.5	91.2	93.5	91.8
Maize	46.9	86.1	35.0	100.0	74.5
Barley	-	100.0	100.0	90.0	94.3
Sugarcane	-	-	59.8	100.0	82.6
Gram	-	-	-	100.0	100.0
Masoor	100.0	100.0	-	100.0	100.0
Mustard	-	72.7	92.3	96.6	94.8
Other oilseeds	-	100.0	-	100.0	100.0
Potato	100.0	94.9	100.0	90.2	91.2
Onion	-	100.0	-	-	100.0
Other vegetables	100.0	65.2	95.0	100.0	96.2
Fodder crop	94.6	82.9	81.9	93.2	88.6
Cotton	100.0	100.0	100.0	100.0	100.0
Other crops	-	100.0	-	87.5	88.2
All crops	87.7	89.8	92.3	95.8	94.1
Bihar					
Paddy	91.9	68.4	93.0	90.6	85.2
Wheat	99.7	93.8	99.5	100.0	98.4
Maize	92.2	79.4	81.1	89.9	86.5
Barley	100.0	100.0	-	-	100.0
Other millet	100.0	5.1	94.2	96.4	88.2
Sugarcane	100.0	100.0	100.0	-	100.0
Gram	100.0	60.4	96.3	72.9	84.3
Arhar	64.9	100.0	91.7	100.0	91.3
Khesari	98.0	99.9	93.9	78.7	91.0
Masoor	99.9	92.0	92.1	100.0	95.3
Other pulses	100.0	51.1	99.8	100.0	87.4
Mustard	100.0	100.0	96.3	100.0	98.8
Other oilseeds	100.0	100.0	100.0	100.0	100.0
Potato	98.6	100.0	100.0	100.0	99.7
Onion	100.0	100.0	100.0	100.0	100.0
Other vegetables	100.0	100.0	100.0	81.4	91.1
Other crops	100.0	48.9	100.0	83.9	75.8
All crops	95.3	81.3	93.5	93.0	90.8

Source: Field Survey.

Table 3.11

Proportion of Area under HYVs

Crop	Marginal	Small	Medium	Large	Total
Punjab					
Paddy	100.00	98.97	100.00	99.20	99.33
Wheat	85.48	89.97	94.39	95.35	94.01
Maize	31.25	81.01	35.00	89.47	65.83
Barley	0.00	100.00	100.00	100.00	100.00
Gram	0.00	0.00	0.00	100.00	100.00
Masoor	0.00	0.00	0.00	100.00	63.79
Mustard	0.00	75.76	92.31	100.00	97.86
Other oilseeds	0.00	54.35	0.00	100.00	74.85
Cotton	100.00	100.00	100.00	98.89	99.14
Sugarcane	0.00	0.00	100.00	100.00	100.00
Potato	100.00	99.36	100.00	95.72	96.23
Onion	0.00	57.14	0.00	0.00	57.14
Other vegetables	100.00	65.22	85.00	79.70	81.02
Fodder crop	74.50	57.52	64.85	73.20	68.04
Other crops	0.00	100.00	0.00	62.50	64.71
All crops	79.84	85.94	91.10	95.16	92.66
Bihar					
Paddy	78.69	54.55	58.51	70.55	63.99
Wheat	88.85	69.70	79.23	79.29	79.02
Maize	70.39	58.70	57.61	64.47	63.10
Barley	0.00	0.00	0.00	0.00	0.00
Other millet	96.45	0.00	4.71	0.00	10.59
Gram	0.00	0.00	2.31	57.05	22.23
Arhar	28.07	77.05	20.53	24.75	30.93
Khesari	17.41	9.83	11.13	13.30	12.18
Masoor	24.35	13.60	8.42	25.47	16.05
Other pulses	54.49	4.89	16.64	0.00	14.09
Mustard	77.40	36.76	55.51	89.54	66.95
Other oilseeds	100.00	66.67	100.00	73.75	75.11
Sugarcane	45.12	36.17	42.80	0.00	41.13
Potato	98.60	83.17	84.27	100.00	90.65
Onion	0.00	100.00	66.30	100.00	82.81
Other vegetables	100.00	35.98	0.00	21.80	49.10
Other crops	0.00	51.12	100.00	76.83	69.23
All crops	76.17	55.35	53.65	65.04	61.55

Source: Field Survey.

Table 3.12

Sources of Information on Seed and Technology

(Per cent of households)

Category	*Agricultural University*	*Private Companies*	*Seed Vendors*	*Other Farmers*	*Any Other*
Punjab					
Source of Information					
Marginal	2.4	1.2	44.4	50.9	1.2
Small	3.2	1.8	46.3	48.6	0.0
Medium	3.2	1.6	42.9	51.9	0.5
Large	2.9	2.6	46.6	45.3	2.6
Total	2.9	1.9	45.3	48.6	1.2
Source of Technology Provider					
Marginal	5.3	12.0	44.4	37.6	0.8
Small	5.3	15.4	43.1	36.2	0.0
Medium	8.9	12.7	36.1	42.4	0.0
Large	4.7	13.3	42.6	39.5	0.0
Total	5.9	13.5	41.6	38.9	0.1
Bihar					
Source of Information					
Marginal	0.3	1.9	34.8	60.4	2.7
Small	0.4	1.7	40.9	53.6	3.4
Medium	0.0	0.6	41.1	54.4	3.8
Large	0.0	5.4	52.7	39.3	2.7
Total	0.2	2.0	39.9	54.8	3.1
Source of Technology Provider					
Marginal	0.6	1.4	36.5	56.0	5.5
Small	0.0	2.3	43.8	49.8	4.1
Medium	0.0	0.7	55.6	38.4	5.3
Large	1.0	2.9	58.8	33.3	3.9
Total	0.4	1.7	44.8	48.3	4.9

Source: Field Survey.

Adoption of any new technique by the farmers mainly depends on dissemination of information about the techniques up to the grassroot level, and the availability of the technique to the farmers at a reasonable price. Table 3.12 shows the sources of information as well as seed and technology providers to the selected households in the two states. It is evident from the data that seed vendors and neighbouring farmers were the main sources of information to the farmers in both the states. Around 45 per cent of the farmers in Punjab and 40 per cent in Bihar received information about the new seeds (and other techniques) from the seed vendors. Another 49 and 55 per cent farmers in Punjab and Bihar respectively, depended on their neighbouring farmers for such information.[3] Only 5 per cent of the farmers in Punjab and 2 per cent in Bihar depended on more reliable sources of information, namely the agricultural university in the state or private seed companies, for such information. This reflects apathy on the part of state agencies and ignorance on the part of farmers. It also depicts lack of interest by the agricultural universities in the states/ country in involving the farming community in the research activities by going beyond the laboratory. Similarly, a vast majority of the farmers (around 80 to 90 per cent) depended on seed vendors and neighbouring farmers for getting access to these new seeds and other components of technology in both the states. In Punjab, in addition to seed vendors, seed companies were also playing a major role by providing their new inventions directly to the farmers. Around 14 per cent of our selected farmers from Punjab obtained their seeds etc., directly from the companies at a reasonable price. In Bihar, on the other hand, there was hardly any company providing these services to the farmers directly, and farmers accessed the new seeds and other technological inventions only through the vendors of the companies. In Punjab, 6 per cent of the farmers obtained their seeds from the agricultural university (Punjab Agricultural University and its outlets), whereas in Bihar the corresponding number was less than 1 per cent.

Cropping Pattern

Much has already been said about the prevalence of the wheat and rice cycle in Punjab. The perpetual dependence on these two crops and high intensity of water use by rice has led to land and water degradation and environmental externalities which have reached such a proportion that it has started threatening the sustainability of agriculture in the state. The Johl Committee (1986 and also 2002) recommended that at least 20 per cent of the area presently under wheat and paddy must be replaced by other

---

3. These data also tally with the results of 59th Round of NSSO.

competitive and profitable alternative crops, like fruits and vegetables. To give a boost to the diversification of agriculture in the state, the Punjab Government launched a multi-crop, multi-year contract farming scheme in 2002. The Punjab government through PAFC chalked out a very ambitious plan of bringing 2.5 million acres of rice-wheat area under alternative crops through contract farming by 2007 (Kumar, 2005). Although the Punjab government has stepped in very seriously to bring about the desired changes in the cropping pattern in the state, a lot still has to change on the crop and output diversification fronts.

With the signing of the General Agreement on Tarrifs and Trade (GATT) in 1994, a lot has changed on the domestic and global front. In order to capture the impact of these changes on the agriculture sector, we have tried to incorporate the broad changes that have taken place in the cropping pattern in the two states under study during this period. The trends of cropping pattern in the two states during the phase of reforms are presented in the Tables 3.13 and 3.14. It is seen from the data that despite all efforts of the state government, wheat and rice still completely dominate the cropping pattern among the selected households in Punjab. In fact, the area under paddy has gone up from 28 per cent before the process of reforms started in 1995, to 33 per cent in the post-reforms period. The area under wheat has come down slightly during this period. These results verify the findings by another study on Punjab (Kumar, 2005), which concludes that due to the efforts of the Punjab government some diversification in the *rabi* season from wheat to other vegetables, oilseeds, combined cropping has taken place. While in the *kharif* season no such lucrative alternatives were available to the farmers as *kharif* pulses and oilseeds, the possible alternatives could not compete with paddy in terms of profitability.

Among other major crops namely, potato, fodder and cotton, and other minor crops like coarse grains and pulses, there was hardly any change in the area under cultivation during the liberalisation period in Punjab. Comparing the farm size categories, it is observed that marginal farmers devoted comparatively less area to paddy than the large farmers. As paddy needs assured irrigation and in the event of erratic electric supply as pointed out by a number of earlier studies, farmers depend on alternate diesel pump sets and generators for lifting groundwater for paddy. As these alternate sources were expensive, a large number of marginal farmers opted for maize in the *kharif* season despite its low profitability. Table 3.13 shows that the share of maize in gross cropped area was above 11 per cent in the case of marginal farmers, while it was only 1 per cent in the case of large

Table 3.13

Changes in Cropping Pattern during Reform Phase in Punjab

(Per cent of GCA)

Crop	*Marginal*	*Small*	*Medium*	*Large*	*Total*
Before 1995					
Paddy	18.63	25.26	32.40	28.05	27.81
Wheat	41.47	42.93	41.01	39.68	40.44
Maize	10.42	4.76	1.19	1.35	2.32
Barley	0.00	0.52	0.15	0.33	0.31
Gram	0.00	0.00	0.00	0.07	0.05
Masur	0.34	0.00	0.00	0.07	0.07
Mustard	0.00	0.52	1.11	0.69	0.70
Other oilseeds	0.00	0.17	0.00	0.02	0.04
Sugarcane	0.00	0.00	0.64	0.20	0.23
Cotton	2.20	1.04	4.60	5.15	4.33
Potato	1.60	5.49	4.08	15.22	11.32
Other vegetables	0.80	0.99	2.08	1.10	1.22
Fodder crop	24.53	18.22	12.75	7.82	11.00
Other crops	0.00	0.09	0.00	0.26	0.18
After 1995					
Paddy	22.81	27.14	32.25	35.14	33.12
Wheat	34.32	38.59	39.13	33.72	35.21
Maize	11.77	5.52	2.13	1.22	2.38
Barley	0.00	0.42	0.48	0.26	0.30
Gram	0.00	0.00	0.00	0.05	0.03
Masur	0.31	0.03	0.00	0.05	0.05
Mustard	0.00	0.58	0.69	1.53	1.20
Other oilseeds	0.00	0.26	0.00	0.04	0.06
Cotton	1.29	1.68	4.05	4.61	4.00
Sugarcane	0.00	0.00	0.45	0.14	0.17
Potato	1.47	5.47	4.60	14.93	11.48
Onion	0.00	0.41	0.00	0.00	0.05
Other vegetables	1.47	0.80	2.13	1.26	1.35
Fodder crops	26.56	19.03	14.08	6.87	10.44
Other crops	0.00	0.07	0.00	0.20	0.15

Source: Field Survey.

farmers. Similar startling differences among small and large farmers were observed in the case of potato and fodder crops. Whereas marginal farmers planted only 1 per cent area under potato, large farmers planted potato on more than 14 per cent of the area. As against this, large farmers had only 7 per cent area under fodder crops while marginal farmers devoted 27 per cent of their area to such crops in post-reform period. The logic for higher area being devoted to potato by large farmers is simple, as potato—a vegetable crop—involves a huge amount of expenditure on seed and other inputs. Also the crop is very risky in terms of net returns as both yield and price of the produce fluctuate a lot. Probably marginal and small farmers falling short of funds generally avoid taking such risks. In the case of fodder crops, it is evident that marginal and small farmers try to supplement their farm earnings through livestock rearing. Therefore, they devoted comparatively higher area to fodder compared to medium and large farmers.

Table 3.14 presents the cropping pattern in Bihar before and after 1995, the period that corresponds to the ongoing reform process in the Indian economy. It is apparent from the data that around 90 per cent of the total cropped area in Bihar was under food grains. Paddy and wheat were the dominant food crops in Bihar as was also the case in Punjab. However, unlike Punjab where cotton was the second leading *kharif* crop, maize occupied a principal place among *kharif* crops in Bihar. In the *rabi* season, apart from wheat, *khesari* and *masoor*, the two pulse crops, occupied a significant area of around 12 per cent in the gross cropped area. Mustard, sugarcane and a few vegetables crops like potato and onion were the other minor crops grown in Bihar. Comparing the two periods before and after 1995, apparently liberalisation has not touched the rural areas in Bihar, as no major change is visible in the cropping pattern of the state during the last decade. Almost same percentage of area was devoted to wheat and paddy before 1995 as it is at present. Among other major crops like maize, *masoor* and *khesari*, no major difference appears in the percentage of area under them during this period.

Comparing farm size categories, it is seen from the table that marginal and small farmers planted more area under paddy compared to large farmers, a trend that was completely opposite to Punjab. It was seen in the previous chapter that more than 90 per cent holdings were classified

Table 3.14

Changes in Cropping Pattern during Reform Phase in Bihar

(Per cent of GCA)

Crop	*Marginal*	*Small*	*Medium*	*Large*	*Total*
Before 1995					
Paddy	33.66	38.69	34.48	29.38	33.51
Wheat	33.92	28.39	23.47	27.77	27.60
Maize	21.35	16.95	12.59	21.11	17.74
Barley	0.03	0.06	0.00	0.00	0.02
Other millet	0.47	0.26	1.48	0.14	0.62
Gram	0.52	0.60	1.44	1.86	1.26
Arhar	0.27	2.03	1.20	0.93	1.13
Khesari	2.70	3.12	5.96	5.00	4.53
Masoor	3.22	3.67	13.27	5.82	7.19
Other pulses	0.40	0.19	0.47	0.40	0.38
Mustard	0.68	0.85	1.02	2.46	1.41
Other oilseeds	0.12	0.28	0.00	0.83	0.35
Sugarcane	0.09	1.91	2.76	0.00	1.23
Potato	1.52	1.19	1.30	0.56	1.07
Onion	0.14	0.55	0.42	0.46	0.41
Other vegetables	0.78	1.06	0.02	2.05	1.03
Other crops	0.12	0.21	0.12	1.21	0.50
After 1995					
Paddy	33.37	40.92	33.79	27.28	33.39
Wheat	32.45	28.20	25.69	28.05	28.20
Maize	20.46	15.14	13.81	24.50	18.56
Barley	0.11	0.19	0.00	0.00	0.06
Other millet	0.29	0.22	1.37	0.50	0.64
Gram	0.47	0.47	1.72	1.33	1.09
Arhar	0.58	0.48	1.25	0.61	0.75
Khesari	2.70	3.99	6.35	4.51	4.59
Masoor	4.29	3.77	9.17	5.44	5.91
Other pulses	0.34	0.48	0.49	0.42	0.44
Mustard	0.66	1.07	1.35	1.55	1.22
Other oil seeds	0.13	0.24	0.06	1.37	0.51
Sugarcane	0.17	1.34	3.03	0.00	1.20
Potato	1.45	1.09	1.24	0.73	1.09
Onion	0.06	0.41	0.47	0.28	0.33
Other vegetables	2.38	1.23	0.02	2.17	1.38
Other crops	0.10	0.77	0.18	1.28	0.64

Note: GCA - Gross cropped area.
Source: Field Survey.

as marginal and small farmers in Bihar. As subsistence concerns dominate among these farmers and rice is the staple food in Bihar, these farmers grew rice mostly for self-consumption. Moreover, the amount of rainfall in Bihar is much higher than in Punjab and paddy is grown mostly in heavy rainfall areas. In Bihar, farmers are not as dependent on artificial irrigation as they are in Punjab. Another noticeable point about cropping pattern is that green fodder was a major crop at least among the marginal and small farmers in Punjab. However, in Bihar green fodder was not grown to feed the animals as dry fodder straw of wheat, paddy and other crops and grazing land was generally used to feed the animals. As regards other crops, no major differences were observed across farm size in Bihar.

Economics of Production and Resource Use Efficiency

Gross farm output is a function of area sown and yield rate per unit of area, which in turn depends on land availability, technology used and composition of farm inputs used by the farmers. Table 3.15 presents the value of gross output per household, per acre and per capita among the selected households in Punjab and Bihar. A positive relation between the

Table 3.15

Gross Value of Output (Main plus Byproduct)

(in Rs)

Category	*Output per Household*	*Output per Acre*	*Output per Capita*
Punjab			
Marginal	36108	22556	6933
Small	99701	25169	15224
Medium	197763	25665	30587
Large	658875	28864	78658
Total	259703	27662	38731
Bihar			
Marginal	18507	15764	2554
Small	43143	11717	4722
Medium	82410	11331	7481
Large	171045	10657	14295
Total	48320	11831	5514

Source: Field Survey.

value of output per household and the farm size is clearly visible for both Punjab as well as Bihar. A positive relation is not surprising as it indicates the effect of increasing land across farm size. However, it is interesting to note that the average value of output per household in Punjab was almost five times higher than that of Bihar. Comparing the farm size categories, one observes that the difference between Punjab and Bihar was not as high among marginal and small farmers as it was among the large farmers. The explanation for this lies in the obvious fact that difference in land operated was much higher among large farmers than among marginal and small farmers, which is also visible in value of output per household.

Nonetheless, the household analysis does not reveal the whole picture as it ignores the differences that arise due to differences in area operated by the households. Therefore, to make the picture clear, productivity comparisons across farm size are presented. The productivity statistics in terms of value of output per acre given in the table clearly shows that farming in Punjab stands way ahead of Bihar. The average productivity per acre in Punjab was Rs 27,662, which was more than twice the corresponding figure of Bihar, Rs 11,831. The difference in productivity between the two states was much less in the case of marginal farmers, Rs 6,792 which increased to Rs 18,207 in the case of large farmers. Comparison across the size classes in the two states further presents very interesting trends. While looking at the value of output per acre in Punjab, a positive relation between productivity and farm size was clearly evident from the data. However, the trends were completely opposite in the case of Bihar. In the latter case, value of output per acre decreased as the farm size increased, establishing a negative relationship between the two. Thus, a seemingly positive relation between farm size and productivity in Punjab and a negative relationship in Bihar gives rebirth to the age-old debate on the inverse farm size-productivity relationship in the country. The explanation for such a relationship could also be sought in the literature.[4] As Punjab agriculture has completely metamorphosed into mechanical

4. On farm size productivity relationship two different opinions are given in the literature. The first opinion given by Sen (1964), Khusro (1964), Bhardwaj (1974) and Rao (1965) argued that as the size of holdings increase, productivity declines. The inverse relationship between farm size and productivity has been questioned by the second group of economists like Rudra (1968), Rao (1967) and others. For details see Kumar (1991).

farming, the traditional advantage of higher input intensity that operated in the favour of small and marginal farmers has completely ceased, and the trends of economies of scale on large farming have turned the productivity advantage in the favour of large farmers. Bihar agriculture seems to be caught in the web of subsistence farming with a very high percentage of holdings being marginal and small. Therefore, mechanisation and commercialisation of agriculture in the state is still at a primitive stage, tilting the advantage of input intensity in favour of small farmers and resulting in higher productivity for them. The trends in output per capita were similar to the trends observed in the case of output per household in both the states.

The productivity or value of output presents the revenue earned out of farming operations. In order to see the returns from the farming business, it is essential to take a note of operational and other fixed costs incurred by the farmers. Farming in Punjab can be termed commercial farming, which also means higher dependence on the market for various inputs. Since it involves higher expenses to the farmers, it might turn out that the farm business income or net returns from farming might not be as lucrative in Punjab compared to Bihar, as it appears to be from their crop productivity.

The cost of cultivation per household for both the states is presented in Table 3.16. The cost of cultivation includes all the elements of input costs involved in the production of crops, right from the stage of preparatory tillage to the final stage of collecting produce in the form of grains and their by products. A glance at the cost statistics reveals very interesting results for Punjab and Bihar. Comparing the percentage expenditure on each input, it is seen from the table that in Punjab, the highest expenses were incurred on manure, chemical fertilisers and pesticides. These items together constituted around 29 per cent of total cost of cultivation. A large number of scholars have already expressed grave concern on increasing level of degradation of soils caused by excess use of chemicals like fertiliser, pesticides and herbicides in Punjab. These results further verify the intensification of chemical use in the state.

Table 3.16

Input Use and Cost of Cultivation

(Rs per household)

Crop	*Marginal*	*Small*	*Medium*	*Large*	*Total*
Punjab					
Seed	1407	3648	8322	18777	8232
	(8.4)	(7.5)	(8.8)	(7.4)	(7.7)
Manure/fertiliser/pesticides	3544	10776	24069	79570	30847
	(21.1)	(22.2)	(25.4)	(31.2)	(28.7)
Irrigation	391	399	416(	725	492
	(2.3)	(0.8)	0.4)	(0.3)	(0.5)
Hired tractor/	2379	4951	5767	13685	6910
bullock maintenance	(14.2)	(10.2)	(6.1)	(5.4)	(6.4)
Own tractor charges	933	5478	14550	47155	17792
	(5.6)	(11.3)	(15.4)	(18.5)	(16.6)
Threshing/combine charges	544	1474	2322	5166	2439
	(3.2)	(3.0)	(2.5)	(2.0)	(2.3)
Total material cost	9197	26727	55445	165078	66713
	(54.8)	(55.2)	(58.5)	(64.8)	(62.1)
Hired labour charges	1885	5428	10733	31354	12841
	(11.2)	(11.2)	(11.3)	(12.3)	(12.0)
Imputed cost of	4248	6639	8587	10031	7358
family labour	(25.3)	(13.7)	(9.1)	(3.9)	(6.9)
Rent paid for	1467	9661	19974	48476	20477
leased-in land	(8.7)	(19.9)	(21.1)	(19.0)	(19.1)
Total cost	16798	48455	94739	254939	107389
	(100)	(100)	(100)	(100)	(100)
Bihar					
Seed	808	1752	3114	8183	2051
	(11.2)	(11.6)	(12.3)	(16.6)	(13.0)
Manure/fertiliser/pesticides	312	826	2003	5411	1153
Irrigation	(4.3)	(5.5)	(7.9)	(11.0)	(7.3)
	504	734	868	721	643
	(7.0)	(4.9)	(3.4)	(1.5)	(4.1)
Hired tractor/	63	153	499	652	210
bullock maintenance	(0.9)	(1.0)	(2.0)	(1.3)	(1.3)
Own tractor charges	280	1171	2121	7725	1434
	(3.9)	(7.8)	(8.4)	(15.7)	(9.1)
Threshing/combine charges	512	1054	2018	1828	1018
Total material cost	(7.1)	(7.0)	(8.0)	(3.7)	(6.5)
	2479	5690	10624	24519	6509
	(34.3)	(37.8)	(42.0)	(49.9)	(41.3)
Hired labour charges	903	3088(	7715	16071	3882
	(12.5)	20.5)	(30.5)	(32.7)	(24.6)
Imputed value of	2668	4423	5830	6891	4004
family labour	(37.0)	(29.4)	(23.0)	(14.0)	(25.4)
Rent paid for	1170	1858	1150	1678	1379
leased-in land	(16.2)	(12.3)	(4.5)	(3.4)	(8.7)
Total cost	7220	15059	25319	49160	15774
	(100.0)	(100.0)	(100.0)	(100.0)	(100.0)

Note: Figures in parentheses are respective percentages of total cost.

Source: Field Survey.

Table 3.17

Input Use, Output and Returns Realised

(Rs per acre)

	Marginal	*Small*	*Medium*	*Large*	*Total*
Punjab					
Seed	879	921	1080	823	877
Manure/fertiliser/pesticides	2214	2720	3124	3486	3286
Irrigation	244	101	54	32	52
Hired tractor/bullock maintenance	1486	1250	748	600	736
Own tractor charges	583	1383	1888	2066	1895
Threshing charges	340	372	301	226	260
Total material cost	5745	6747	7195	7232	7106
Hired labour charges	1178	1370	1393	1374	1368
Total family labour (man-days)	33	21	14	5	10
Imputed cost of family labour	2654	1676	1114	439	784
Rent paid for leased-in land	916	2439	2592	2124	2181
Total cost	10493	12232	12295	11168	11438
Output	22556	25169	25665	28864	27662
Farm business income (FBI)	12063	12937	13370	17696	16224
Output-input ratio	2.15	2.06	2.09	2.58	2.42
Bihar					
Seed	688	476	428	510	502
Manure/fertiliser/pesticides	265	224	275	337	282
Irrigation	430	199	119	45	157
Hired tractor/bullock maintenance	54	41	69	41	51
Own tractor charges	238	318	292	481	351
Threshing charges	436	286	277	114	249
Total material cost	2112	1545	1461	1528	1594
Hired labour charges	769	839	1061	1001	950
Total family labour (man-days)	45	24	16	9	20
Imputed value of family labour	2273	1201	802	429	980
Rent paid for leased-in land	997	505	158	105	338
Total cost	6150	4090	3481	3063	3862
Value of output	15764	11717	11331	10657	11831
Farm business income (FBI)	9614	7627	7850	7594	7968
Output-input ratio	2.56	2.86	3.26	3.48	3.06

Source: Field Survey.

Expenses on use of machines or mechanical inputs like hired tractors expenses or maintenance-expenses of own tractors, was the next component in the total cost of cultivation. This particular component of hired/owned tractor constituted around 23 per cent of total cost to the farmers. The rental paid for leasing-in land was fairly high and it constituted the third major item in the cost of cultivation, 19 per cent of total cost. Expenses incurred on human labour, including the imputed value of family labour were the other important component of cost of cultivation. Because of the subsidy involved in canal water and electricity used for agricultural purposes, the irrigation charges were less significant in total cost of cultivation in Punjab as is evident from the table. Comparing cost percentage across different farm size categories, it is observed from the table that hired tractors and family labour constituted higher costs to the marginal farmers whereas, own tractor expenses and expenses for leased-in land was comparatively higher in the case of large farmers.

The percentage disbursement of cost among different inputs in Bihar presents a contrasting picture. Labour (hired plus imputed value of family labour) is the single largest item, which constituted around 50 per cent of total cost of cultivation in Bihar. The components of modern technology namely, fertiliser-pesticides, irrigation and tractors together accounted for around 22 per cent of total cost of cultivation in Bihar, whereas, their contribution was more than 52 per cent in the case of Punjab. These figures verify our earlier assertion that Punjab agriculture is largely mechanised and has become commercial farming, whereas Bihar agriculture is still subsistence-oriented, overburdened with family labour, especially in the case of small and marginal farmers who constitute 90 per cent of total holdings in the state.

Production Function

In the previous section we discussed productivity, cost and returns from farming in Bihar and Punjab. However, the analysis presented in the tables has the limitation of not being able to segregate the aggregate effect of multiple variables present in these equations. To know the direction as well as the quantum effect of each variable present in a relationship, we prefer multiple regression analysis. The ordinary least squares (OLS) estimates of farm productivity (Cobb-Douglas) in both these states are presented in Table 3.18. The variables considered for explaining the productivity variations include: Farm size (net operated area); seed; manure, fertiliser and plant protection chemicals; tractor and other machines used in the sowing and harvesting operations; canal and tube well/pump set (electric or diesel) used for irrigation; and total human labour (family and hired) used in the agriculture operations. All the variables were in value terms except for human labour, which was in value terms for hired labour and number of days for total labour (family plus hired). The value of each variable was calculated using the actual prices paid by the farmer at the time of the farm operations. The production function was estimated in terms of value of output per acre, so all variables were divided by the operated area to get value per unit of area. The relationship was preferred in proportionate form as in absolute terms most of the regressors were highly collinear.

In the case of Punjab, three separate regressions were used with three different irrigation variables. The first irrigation variable was defined in terms of electric tubewells with submersible pumps while in the second variable tube wells with non-submersible pumps were used to represent the irrigation variable. The irrigation variable was taken as the sum of the expenditure on electric pump sets (submersible plus non-submersible), diesel pump sets and the amount paid for using canal water during the year, in the third case. In Bihar, however, in almost all cases, no electric tube well was used. Therefore, the variable representing irrigation included the expenditure incurred on fuel used to run the diesel pump sets and the amount paid for the canal water used by the cultivators. In both the states, the labour variable was also defined in two different ways. In the first case, expenditure incurred on hired labour (permanent or casual) was used as an independent variable, while a second variable, the total numbers of days (family plus hired) was taken as the autonomous variable.

In Punjab, except seeds, all other variables emerge positive and statistically significant. The significant and positive coefficient of operated area indicates that farm productivity increased with the rise in farm size. Along with this implicit inference of the positive relationship is the inference that the large operators were comparatively more efficient, as they generated higher output per acre of net sown area compared to the small size holders. The positive relationship in Punjab seems to be an outcome of the high use of mechanical operations in the sowing and harvesting activities in the state, as the latter leads to a clear advantage of economies of scale to large holders. The importance of mechanical operations was also underlined by the high elasticity of mechanisation and tractors. Fertiliser and plant protection and human labour were the other variables, which had high elasticity.

Table 3.18

Results of the Production Function Exercise

Dependent Variable: Log Value of Output per Acre

Independent Variables	Punjab				Bihar	
Constant	5.27 (17.0)	5.07 (16.1)	5.22 (16.6)	5.16 (16.1)	6.00 (22.3)	6.33 (22.3)
Log net operated area (Per acre)	0.04 (2.3)	0.06 (3.1)	0.05 (2.5)	0.01 (0.5)	0.01 (0.6)	-0.08 (-3.4)
Log seed (Rs per acre)	0.02 (1.1)	0.02 (1.1)	0.03 (1.3)	0.04 (1.8)	0.01 (0.9)	0.02 (2.0)
Log (fertiliser + plant protection) (Rs per acre)	0.28 (6.7)	0.28 (6.6)	0.27 (6.5)	0.31 (7.3)	0.1 (4.4)	0.29 (6.8)
Log Mechanisation & tractor (Rs per acre)	0.22 (6.4)	0.24 (6.9)	0.22 (6.2)	0.22 (5.9)	0.13 (3.6)	0.12 (3.1)
Log labour days (family +hired) (No. of days per acre)	0.17 (3.7)	0.18 (3.7)	0.18 (3.8)		0.28 (7.1)	
Log hired labour expenditure (Rs per acre)				0.07 (2.8)		0.03 (2.6)
Log pump set – diesel (Rs per acre)					-0.001 (-0.2)	0.01 (0.9)
Log submersible pump (Rs per acre)	0.020 (4.2)			0.020 (4.1)		
Log non-submersible pump (Rs per acre)		0.004 (0.8)				
Log pump sets (diesel + electric) (Rs per acre)			0.016 (3.1)			
Adjusted R-squared	0.53	0.50	0.52	0.52	0.39	0.31
F-statistic	56.8*	50.9*	54.0*	51.3*	37.4*	26.8*
No of observations	296	296	296	284	351	351

Note: The figures in parentheses are respective 't' values.
* Significant at 1 per cent level, ** Significant at 5 per cent level.

Source: Field Survey.

In the case of irrigation, it is evident from the results that submersible pump sets impacted productivity positively, as its coefficient was significant, while the coefficient of non-submersible pumpsets was insignificant. The elasticity of submersible tubewells indicated that a 10 per cent investment by farmers in the submersible tubewells led to 0.2 per cent hike in land productivity, whereas investment in non-submersible pumps did not have any significant impact on the farm productivity. In the case of human labour, the high elasticity indicated that despite large mechanisation of agriculture in Punjab, there was still scope for manual labour, and in a few operations in agriculture, like preparation of the field, transplanting, caring, fodder cutting etc., labour could not be replaced by the machines.

In Bihar, the variable representing the operated area did not emerge statistically significant in the first equation, while it was significant with a negative sign in the second equation. The negative sign of the coefficient of farm size emphasises the fact that small farmers were more efficient in Bihar, as they produced higher per acre output compared to large farmers. The explanation of the dichotomy of higher productivity for large farmers in Punjab against that for small farmers in Bihar lies in the fact that the advantage of economies of scale supported by large scale mechanisation went in favour of large farmers in Punjab, while the advantage of higher intensity of labour input supported by family labour went in favour of small farmers in Bihar (see Table 3.17).

Among the explanatory variables in Bihar, plant protection, mechanisation and human labour were the other significant variables with a positive sign. The elasticity of the variable representing mechanisation was almost half that for Punjab, indicating lower productivity of machine labour in Bihar, because of the predominance of small farming in the state. On the other hand, the elasticity of human labour was almost one and a half times higher than in Punjab, indicating higher productivity for manual agriculture in Bihar, supported by the fact that the average size of holdings in the state was only 1.8 acres, and 90 per cent holdings belonged to the small and marginal size. Last but not the least, the irrigation variable was not significant, emphasising the predominance of rainfed agriculture in Bihar.

To sum up the functional analysis, the coefficient of operated area was significant with opposite signs in the two states. Its positive sign in Punjab indicated higher productivity for large farmers, while the negative sign in Bihar emphasised higher efficiency of small farmers in terms of higher land productivity. The increasing productivity with increasing farm size in Punjab was the outcome of higher use of mechanical operations, while higher productivity for small farm size in Bihar was the result of small-scale manual farming. Fertilisers, mechanisation and human labour were the other significant variables in the two states. The elasticity of mechanisation in Bihar was almost half that of Punjab indicating lower productivity of machines in Bihar because of predominance of small farming in the state. On the other hand, the elasticity of human labour in Bihar was almost one and a half times higher than in Punjab indicating higher productivity for manual agriculture in that state. In the case of irrigation, submersible pumpsets contributed significantly to land productivity, while investment in non-submersible tubewells failed to raise the land productivity in Punjab. In Bihar, however, only diesel pumpsets were used which also remained insignificant in raising land productivity.

4 Market and Prices

The structure of agriculture produce marketing in India varies across different commodities, while at the aggregate level it consists of a mix of public and private sectors. Barring direct intervention by the government in some commodities, marketing in most others have remained dominated by the private sector. According to Acharya (1994), the quantity of agricultural produce handled by government agencies has not been more than 10 per cent of the total value of marketed surplus. Another 10 per cent of the marketed surplus is handled by the cooperatives. Thus, the remaining 80 per cent of the marketed surplus comes within the ambit of private trade.

Creation of regulated markets under the provisions of the 'Agricultural Produce Markets Act' was a major step in improving agricultural marketing in India. These regulated markets function with clear rules and regulations with regard to open auctioning and fixed marketing charges, including those for various operations. These markets generally provide adequate infrastructure in terms of marketing yards and succeeded in reducing many illegal exactions earlier charged by the traders. With the setting up of regulated *mandis*, market sales by farmers have increased and physical losses during handling, storage and transportation have been reduced, along with the rationalisation of market charges. The process of price settlement in most of the markets has become quite transparent, and backward and forward linkages of wholesale *mandis* have been considerably strengthened. Owing to a widening of the production base of the agricultural sector, the market orientation of the farm sector has considerably increased. The overall impact of the working of these institutions on agricultural marketing has been decidedly positive and has helped to increase its competitiveness and efficiency (Kumar, 1996).

However, these institutional reforms have not been successful in terms of coverage over the whole of India. These regulated markets are generally more successful in agriculturally advanced areas, and have relegated the trader-cum-moneylender to a position of secondary importance, mostly in the areas of the Green Revolution. Market imperfections continue to operate in most of the areas where an agricultural breakthrough has not taken place (Bhalla, 1991). In the backward areas, the number of these markets is limited and the markets continue to be dominated by the trader-cum-moneylender nexus. In the agriculturally advanced regions, the market infrastructure is fairly developed resulting in efficient marketing. In the agriculturally underdeveloped parts of India, it is highly inadequate and consequently, the marketing system continues to be non-competitive and dominated by monopsonistic interests (Kumar, 1996).

As a large part of agricultural produce is marketed through private trade, there are a number of functionaries operating in different areas of marketing of various commodities. There are around two million wholesalers and five million retailers (including four lakh fair price shops operating under the public distribution system, handling the entire trade in food grains in India. Apart from wholesalers and retailers, processors enter the market as bulk buyers and sellers. In the case of rice, there are 91,801 hullers, 4,538 shellers, 8,365 huller-cum-shellers and 34,688 rice mills processing the entire produce of paddy in India (Acharya and Chaudhri, 2001). Wheat processing is dominated by the small flourmills known as *chakkies,* operating at a very small scale in the villages and small towns. According to an estimate, there are around 266,000 such *chakkies* operating at present. In addition to these *chakkies,* which are generally in the unorganised sector, there are around 812 roller flourmills operating at a much larger scale. Around 10.5 million tonnes of wheat are handled by these roller flourmills operating mostly in the private sector. In the case of pulses, there are small and large-scale *dal* mills operating in the organised and unorganised sectors. The estimated number of organised *dal* mills is more than 10,000 while the number of unorganised mills is much higher. In the case of oilseeds and vegetable oil processing, there are around 20,000 expellers, 131,600 *ghanies* and 761 solvent extraction units. However, the processed and semi-processed segment of the total Indian food market accounts for only 25 per cent of the market, and the remaining 75 per cent is constituted by fresh food (Acharya and Chaudhari, 2001). Further, in the case of fruits and vegetables, only 2 per cent of total production is processed and the remaining 98 per cent is traded as fresh farm products in the fruit and vegetable markets.

Against the background of the market structure discussed above, the present chapter makes an attempt to study the marketing behaviour of the selected households in Punjab and Bihar. Sections I and II present an analysis of the marketed surplus by farm size at the aggregate and crop-wise level, respectively. Section III, we try to comprehend the marketing channels followed by the farmers in the two states and review the success of the regulated marketing system. Section IV reviews the cost of marketing incurred by the farmers in the two states. In the section V an attempt is made to compare the prices obtained by farmers in the two states for the major crops, and to observe whether there is any price discrimination prevalent across various farm sizes. A gross price index is calculated for all crops.

Marketed Surplus (Aggregate)

We define market surplus in the present case as the quantity of produce the producer actually sells irrespective of his needs for home consumption and other requirements. In technical terms, this is known as 'marketed surplus'. This term is different from 'marketable surplus' which is defined as the residual left with the producer after meeting his requirements for family consumption, farm needs and payment-in-kind to casual and permanent labour, the landlord, artisans and others (Moore *et al.*, 1973). In the case of the latter, there is a need to ascertain whether the farm household has produced output in excess of all its compulsory retentions. A certain amount of value judgment is involved in the measurement of marketable surplus. We therefore preferred marketed surplus, which does not entail any subjective judgment, as the latter is measured as the volume of output put to sale by the farmer, irrespective of the family requirements.

Table 4.1 presents per household and per acre marketed surplus by the selected farmers in the two states. The data on the percentage of marketed surplus substantiate our earlier argument that Punjab farmers were highly commercialised, and were producing a very high proportion, more than four-fifth of their output, for the market. Bihar farmers on the other hand were subsistence farmers who produced more than half of their output for self-consumption. As in the case of per household output, per household marketed surplus also had a direct relationship with farm size in both the states, as also did per acre marketed surplus in the case of Punjab. In Bihar however, although output had an inverse relationship with farm size, marketed surplus had a direct relationship with farm size. This indicates that despite their lower productivity per acre, large farmers contributed higher output per acre to the market compared to the small and marginal farmers.

Table 4.1

Marketed Surplus by the Selected Households

(Value in Rs)

Category	*Net Output Produced**		*Net Output Marketed*		*Marketed Surplus (%)*
	per Household	*per Acre*	*per Household*	*per Acre*	
Punjab					
Marginal	26510	16560	19796	12366	74.7
Small	80899	20423	67306	16991	83.2
Medium	170407	22114	148236	19237	87.0
Large	608131	26641	512191	22438	84.2
Total	232615	24777	195951	20872	84.2
Bihar					
Marginal	17099	14565	5363	4568	31.4
Small	40276	10938	14741	4004	36.6
Medium	76770	10556	34277	4713	44.6
Large	158654	9885	84705	5278	53.4
Total	44908	10995	19258	4715	42.9

Note: * Net output here excludes the byproducts and green fodder which is generally used to feed own livestock.

Source: Field Survey.

It is observed from Table 4.1 that in Punjab, marginal farmers with 75 per cent marketed surplus competed well with large farmers who contributed 84 per cent of their output to the market. In Bihar, on the other hand, the contribution of marginal farmers was only 31 per cent whereas large farmers contributed around 53 per cent of their output to the market. Table 4.2 further emphasises the higher contribution of large farmers in the market surplus in both the states. It can be seen from the table that the first three categories of farmers together constitute around 72 per cent of holdings and occupy around 32 per cent of area, but contributed only 26 per cent of marketed surplus in Punjab. Large farmers on the other hand, occupy 68 per cent of operated area and contributed 74 per cent share of marketed surplus. In the case of Bihar, the above mentioned three categories constituted around 92 per cent of the holdings that occupied 68 per cent of the area and contributed around 64 per cent to the total marketed surplus. The remaining 36 per cent surplus was contributed by large farmers, who occupied around 32 per cent of the area operated.

Table 4.2

Distribution of Output and Marketed Surplus

	% of Households	*% of Area Aperated*	*% of Output*	*% of Marketed Surplus*
Punjab				
Marginal	25.5	4.3	2.9	2.6
Small	27.2	11.5	9.4	9.3
Medium	19.2	15.8	14.1	14.5
Large	28.1	68.4	73.6	73.6
Bihar				
Marginal	49.4	14.2	18.8	13.8
Small	24.9	22.4	22.3	19.0
Medium	17.5	31.2	29.9	31.2
Large	8.2	32.2	28.9	36.0

Source: Field Survey.

The higher contribution of large farmers to the market surplus is also clear from the comparison of percentage area and market surplus. Studying the data in these two columns carefully, it is evident that in both the states, as farm size increased, percentage of area and marketed surplus also increased, while per cent of area remained higher than per cent of marketed surplus up to the medium size class. In the large size category, percentage of marketed surplus took over the area percentage implying that the large farmers contributed proportionately higher output to the market than their counterparts.

Marketed Surplus (Crop-wise)

Crop-wise comparison of market surplus further confirms the findings that Punjab farming is highly commercialised whereas subsistence predominates farming in Bihar. Table 4.3 presents trends in crop productivity and marketed surplus for these two states. The crop productivity observed in our survey data mostly matches with the productivity trends presented in Chapter 2 for the states based on secondary sources, although in a few cases, productivity among the sample farmers was slightly higher than the state average. Productivity for wheat and paddy, the two major crops grown in Punjab, was 15 and 25 quintals per acre respectively. In Bihar, productivity of paddy was only half that of Punjab, while wheat productivity was two-third of Punjab. Among all major crops, only in the case of maize, did productivity in Bihar exceed that of Punjab.

Table 4.3

Marketed Surplus per Household and per Acre of Cropped Area

(Quintals per acre)

Crop	*Output per Household*	*Output per Acre*	*Marketed Surplus per Household*	*Marketed Surplus per Acre*	*Marketed Surplus as a % of Output*
Punjab					
Wheat	118.0	14.8	90.9	11.4	77.1
Paddy	210.8	24.8	206.1	24.2	97.8
Maize	30.6	7.3	20.3	4.8	66.4
Cotton	49.6	8.3	46.8	7.8	94.5
Mustard	22.8	6.1	16.1	4.4	70.9
Potato	855.2	104.7	555.9	68.1	65.0
Other vegetables	64.2	22.4	58.4	20.3	91.0
Bihar					
Wheat	39.9	10.3	20.3	5.2	51.0
Paddy	133.2	11.5	20.5	1.8	15.4
Maize	47.3	11.8	31.4	7.9	66.4
Gram	6.0	3.4	3.4	1.9	56.4
Arhar	12.2	5.3	6.0	2.6	49.2
Khesari	8.2	3.6	5.0	2.2	61.2
Masoor	9.3	3.7	4.7	1.9	50.2
Mustard	8.3	3.7	2.3	1.0	27.8
Sugarcane	363.5	209.4	326.6	188.1	89.8
Potato	33.2	56.8	18.7	32.0	56.4
Onion	37.0	57.8	29.7	46.3	80.2
Other vegetables	14.2	18.6	4.5	5.9	31.4
Other crops	32.0	20.8	9.4	6.1	29.5

Source: Field Survey.

The last column of Table 4.3 shows the percentage of output marketed for each crop. Wheat and rice were the major staple food crops in Punjab and Bihar, respectively. On an average, 77 per cent of the output of wheat was marketed in Punjab compared to 51 per cent in Bihar. Thus, despite being the staple diet in Punjab, more than three-fourth of the output of wheat was marketed in that state. On the other hand, rice being the staple cereal in Bihar, only 15 per cent of the output was marketed there, compared to 98 per cent of the output marketed in Punjab. This also highlights the dominance of small farmers in Bihar agriculture, where the major staple crop is grown only for home consumption. This constitutes around one-third share in the gross cropped area. Among other major crops, 65 per cent of the production of potato and 95 per cent of cotton was marketed in Punjab. In Bihar, 66 per cent of maize, 50 per cent of *masoor* and 61 per cent of *khesari* was marketed.

Annexure Tables A-4.1 and 4.2 show the percentage distribution of marketed surplus across various farm size categories. Barring a few cases, the percentage of output marketed rose with the rise in farm size in both the states. Comparing the marketed surplus of wheat and paddy, the two major crops occupied around 68 per cent of the gross cropped area in Punjab and 61 per cent area in Bihar. It is seen from the tables that in Punjab, the percentage of output marketed increased from 49 per cent in the case of marginal farmers to 81 per cent in the case of large farmers. The corresponding figures for Bihar were 32 and 65 per cent respectively. The trends were quite interesting in the case of paddy where almost the total output was marketed by small as well as large farmers, (the range of surplus among different size classes being 95.2 to 98.6 per cent) in Punjab, while in Bihar, around 11 per cent output was marketed by marginal, small and medium farmers and 29 per cent by the large farmers. Among commercial crops like cotton and sugarcane, above 90 per cent of the output was marketed by almost all farmers in both the states.

Marketing Channels

Marketing channels through which farmers sell their produce in the market is an important indicator of market efficiency. Marketing of agricultural produce in India is regulated and farmers are directed to sell their produce through regulated *mandis*. However, as was pointed out earlier, regulated *mandis* are operating successfully only in a few states, mostly in the Green Revolution belt. Punjab was the foremost state in introducing agricultural marketing reforms. The Punjab Agricultural Produce Markets Act was enacted in 1939 and the scope of the Act was enlarged in 1961. Agricultural marketing in the state was regularised under this act. These regulated *mandis* functioning under the Agricultural Produce Market Committees (APMCs) are operating successfully in Punjab and a large part of agricultural production is sold through these regulated *mandis*. Bihar seems to be way behind in introducing these reforms. The state is still functioning in the traditional subsistence mode whereby the marketing system is unregulated, and the farmers sell their produce in the village to the shopkeepers, traders, itinerant merchants or the big landlords. This is also evident from our survey results discussed below.

The results presented in Table 4.4 show that in Punjab more than 86 per cent of the output was sold through wholesale markets, i.e., regulated *mandis* functioning under the APMC as discussed above. The other major channel was the local market, which in the case of the two major crops wheat and paddy, also came under the net of regulated *mandis*. The market committees generally set up their centres in the village periphery during the harvest season for procurement of these crops at the minimum support price (MSP) almost at the doorsteps of the farmers. Thus, we can categorise these local markets for wheat and paddy as sub-yards of these regulated *mandis* set up for this special purpose. Only the remaining 2 per cent of the output was sold in the village or through intermediaries in the case of Punjab. The pattern of sale was almost same across various farm sizes in the state.

In Bihar, only 10 per cent of the total marketed surplus was handled through regulated *mandis*. Another 17 per cent of the marketed surplus was handled through local markets, which unlike Punjab, could in no sense be termed as sub-yards of regulated *mandis*. Pre-arranged contract and direct sale to the village were the major channels of sale, which together accounted for around 56 per cent of total market surplus. The pre-arranged contract was not the kind of contract signed between the farmer and agri-business firms as is happening in the case of contract farming in Punjab and many other parts of the country. It was a pre-arranged oral contract between the farmers and either the moneylender or the landlord or

Table 4.4

Marketing Channels through which Product was Sold by the Selected Households

(Rs per household)

	Wholesale Market	Local Market	In the Village (Directly)	Cooperative	Government Agencies	Intermediaries at Farm	Pre-Arranged Contract	Others	Total
Punjab									
Marginal	16074 (81.2)	2275 (11.5)	1447 (7.3)	0 (0.0)	0 (0.0)	0 (0.0)	0 (0.0)	0 (0.0)	19796 (100.0)
Small	56181 (83.5)	8964 (13.3)	1920 (2.9)	0 (0.0)	0 (0.0)	242 (0.4)	0 (0.0)	0 (0.0)	67306 (100.0)
Medium	125589 (84.7)	17147 (11.6)	2146 (1.4)	0 (0.0)	697 (0.5)	2657 (1.8)	0 (0.0)	0 (0.0)	148236 (100.0)
Large	441979 (86.3)	62449 (12.2)	2219 (0.4)	0 (0.0)	0 (0.0)	5015 (1.0)	0 (0.0)	529 (0.1)	512191 (100.0)
Total	167870 (85.7)	23884 (12.2)	1927 (1.0)	0 (0.0)	134 (0.1)	1987 (1.0)	0 (0.0)	149 (0.1)	195951 (100.0)
Bihar									
Marginal	94 (1.8)	596 (11.1)	1845 (34.4)	0 (0.0)	113 (2.1)	427 (8.0)	2073 (38.7)	215 (4.0)	5363 (100.0)
Small	1203 (8.2)	4360 (29.6)	4797 (32.5)	21 (0.1)	295 (2.0)	1359 (9.2)	2503 (17.0)	203 (1.4)	14741 (100.0)
Medium	3673 (10.7)	6172 (18.0)	10914 (31.8)	155 (0.5)	5615 (16.4)	2355 (6.9)	4156 (12.1)	1237 (3.6)	34277 (100.0)
Large	11570 (13.7)	8951 (10.6)	14143 (16.7)	0 (0.0)	0 (0.0)	8397 (9.9)	38531 (45.5)	3114 (3.7)	84705 (100.0)
Total	1937 (10.1)	3193 (16.6)	5175 (26.9)	32 (0.2)	1113 (5.8)	1649 (8.6)	5531 (28.7)	628 (3.3)	19258 (100.0)

Note: Figures in parentheses are percentages of total sale.
Source: Field Survey.

Table 4.5

Marketing Channels through which Product was Sold by the Selected Households

(Rs per acre)

	Wholesale Market	Local Market	In the Village (Directly)	Cooperative	Government Agencies	Intermediaries at Farm	Pre-Arranged Contract	Others	Total
Punjab									
Marginal	10041	1421	904	0	0	0	0	0	12366
Small	14183	2263	485	0	0	61	0	0	16991
Medium	16298	2225	278	0	90	345	0	0	19237
Large	19362	2736	97	0	0	220	0	23	22438
Total	17881	2544	205	0	14	212	0	16	20872
Bihar									
Marginal	80	508	1572	0	96	364	1766	183	4568
Small	327	1184	1303	6	80	369	680	55	4004
Medium	505	849	1501	21	772	324	571	170	4713
Large	721	558	881	0	0	523	2401	194	5278
Total	474	782	1267	8	272	404	1354	154	4715

Source: Field Survey.

sahukar, or any other intermediary. It is apparent that the farmers in Bihar are still suffering due to the nexus of credit-market interlinkages traditionally prevalent all over India. The marginal farmers who contributed a paltry amount to the market surplus suffered much more due to this nexus as the percentage of their sales to the regulated *mandis* was less than 2 per cent.

Marketing Cost

Marketing costs are the expenses incurred in bringing the produce from the farm gate to the consumer. These costs generally include handling, packing and loading charges at the farm place, storage and transportation charges, charges for unloading, cleaning and weighing at the market place and other retail charges to the consumers. However, the present study is concerned with the expenses incurred by the producers in selling their produce to the buyer. How the produce moves into the hands of consumers from the buyer is beyond the purview of this study. Therefore, retail charges are not included in the marketing costs in the present case.

The main items we have included in marketing costs are transportation costs, unloading, cleaning and weighing charges at the market and other charges including the personal expenses incurred by the farmers during their stay in the market to sell their produce. Handling, packing and loading at the farm place are done by the producers themselves and therefore, these items do not count in the marketing costs in the present analysis.

The marketing costs thus computed are presented in Table 4.6. It is rather ironical to observe that the Punjab farmer was paying higher per quintal marketing costs compared to the Bihar farmer. On an average, the Punjab farmer was paying Rs 9 per quintal for selling his produce while the Bihar farmer was paying only Rs 5 per quintal. Moreover, the marketing cost was much higher for a marginal farmer in Punjab compared to large farmer, whereas in Bihar, it was the reverse case. The explanation of this dichotomy lies in the fact that in Bihar, a very high amount of output is sold in the village or to the intermediaries at the farm itself. Therefore, their market cost is partly underestimated. On the other hand, in Punjab, around 98 per cent of the output is sold in the market and therefore farmers have to bear the cost of transportation, market fees and other expenses and thereby end up with higher marketing costs. The advantage of a regulated market system occurs in terms of higher net price for Punjab farmers as compared to Bihar as is highlighted in the next section.

Table 4.6

Marketing Cost of Total Produce Sold

(Rs per quintal)

Category	*Transportation Cost*	*Marketing Fee*	*Any Other*	*Total*
Punjab				
Marginal	9.25	0.98	1.94	*12.18*
	(76.0)	(8.1)	(15.9)	*(100.0)*
Small	8.00	0.46	1.45	*9.91*
	(80.7)	(4.6)	(14.7)	*(100.0)*
Medium	3.87	0.94	1.20	*6.01*
	(64.4)	(15.6)	(20.0)	*(100.0)*
Large	4.59	0.66	4.04	*9.29*
	(49.4)	(7.1)	(43.5)	*(100.0)*
Total	4.92	0.68	3.39	8.99
	(54.7)	(7.6)	(37.7)	(100.0)
Bihar				
Marginal	1.10	0.08	0.56	*1.74*
	(63.1)	(4.5)	(32.3)	*(100.0)*
Small	2.94	0.16	0.33	*3.44*
	(85.6)	(4.8)	(9.6)	*(100.0)*
Medium	7.36	0.01	0.66	*8.03*
	(91.6)	(0.1)	(8.3)	*(100.0)*
Large	1.33	0.34	0.85	*2.53*
	(52.6)	(13.6)	(33.8)	*(100.0)*
Total	4.25	0.13	0.63	5.02
	(84.7)	(2.7)	(12.6)	(100.0)

Note: Figures in parentheses are percentages of total marketing cost
Source: Field Survey.

From the point of view of weightage of different items in total marketing costs, it is seen from the table that transportation costs alone accounted for 55 per cent of total costs in Punjab and 85 per cent in Bihar. The other costs including the cost of storage, spoilage/damage on the field and other expenses accounted for around 38 per cent of total costs in Punjab and around 13 per cent in Bihar. Market fees and other unloading, cleaning and weighing charges were the least, around 8 per cent in Punjab and 3 per cent in Bihar.

Prices Received by the Farmers

The market structure and conduct can best be judged from the pricing efficiency of crops produced and sold by the farmers in the market. A fair deal in the determination of these prices depends upon the availability of market infrastructure and marketing services. Important among these services are availability of well-furnished regulated markets with efficient functionaries; standardised weights and measures; adequate transport and communication facilities; proper storage facilities and a well-connected market information system. In addition, the cultivators would be able to harness the benefits of the marketing network only if they participate in the market with some marketed surplus, after satisfying their home needs.

It was pointed out in the previous section that there was a well-established regulated market (APMC) set up in Punjab, where more than 90 per cent of the output was sold through these regulated *mandis*. In Bihar, on the other hand, the market system was still in shackles. Consequently a very high amount of output was sold in the village or to the intermediaries at the farm itself. Sale through regulated markets was only 10 per cent as discussed earlier. With this given market structure, it is evident from the results of Table 4.7 that gross as well as net prices received by farmers in Punjab for almost all crops were above those received by the farmers in Bihar.

Crop-wise gross and net prices received by the farmers are given in Table 4.7. The net price is the price net of market expenses in terms of charges paid in the market namely, transport, market fees and other expenses incurred by the farmers in selling the produce. Before analysing the prices received by farmers for various crops, it is essential to clarify that the Government of India on the recommendations of CACP (Commission for Agricultural Costs and Prices) announces MSP for 24 major crops, including the major food grains, oilseeds and other commercial crops. These MSPs set the floor price in the market for the respective crops. For a few crops, in addition to announcing the MSP, government (Central or state) also procures output from the farmers through its various agencies like the Food Corporation of India (FCI), National Agricultural Cooperative Marketing Federation of India Ltd. (NAFED), sugar mills, Jute Corporation of India (JCI) and Cotton Corporation of India (CCI), etc.

Table 4.7

Gross and Net Price Received by the Farmers and Marketing Cost of Major Crops

(Rs per quintal)

Crop	*Gross Price*	*Transportation Cost*	*Marketing Fee*	*Other Cost*	*Total Marketing*	*Net Price*
Punjab						
Wheat	628	3.43	1.02	0.17	4.62	623
Paddy	580	6.69	1.15	1.71	9.55	571
Maize	550	7.97	0.00	1.11	9.08	541
Mustard	1643	7.44	0.92	0.00	8.36	1634
Cotton	2139	8.87	0.00	0.90	9.77	2129
Potato	242	3.77	0.12	7.48	11.36	230
Other vegetables	432	14.37	2.21	0.00	16.58	415
Bihar						
Wheat	587	0.78	0.14	0.23	1.15	586
Paddy	452	9.61	0.56	0.56	10.74	441
Maize	390	1.04	0.10	0.69	1.83	388
Gram	1706	0.17	0.00	0.00	0.17	1706
Arhar	1541	0.00	0.00	0.00	0.00	1541
Khesari	929	3.39	0.00	0.00	3.39	926
Masoor	1673	0.02	0.00	0.00	0.02	1673
Mustard	1524	0.00	0.00	0.00	0.00	1524
Sugarcane	69	8.45	0.00	0.09	8.54	61
Potato	306	0.03	0.00	6.20	6.23	299
Onion	451	0.45	0.04	0.00	0.48	450
Other vegetables	383	13.13	2.83	0.00	15.97	367

Source: Field Survey.

It is evident from the table that realised prices of all crops except vegetables were higher in Punjab compared to Bihar. In the case of wheat and paddy, the two major cereals for which procurement was done by the FCI, the realised prices were Rs 628 and Rs 580, respectively in Punjab and Rs 587 and Rs 452, respectively in Bihar. The realised price for wheat in Punjab was almost equal to the MSP price (Rs 630) during the survey year, while the paddy price was above the MSP price, the latter being Rs 550 during the survey year. Thus, these figures clearly indicate that the realised prices in Bihar were much lower than the MSP prices announced by the government for these two crops, which occupied around two-thirds of the cropped area in Bihar. The principal reason for the prices being below the MSP in the case of Bihar seems to be complete absence of procurement operations in the state. During the survey year, the share of Punjab in total procurement was 57 per cent in the case of wheat and 38 per cent in the case of rice, which was highest by any state during that year. On the other hand, no operation was carried out in Bihar in the case of wheat and the share of the state was less than 2 per cent in the case of rice in the total central pool.

Annexure Tables A-4.3 and 4.4 present comparative analyses of realised prices by various farm size categories. It is seen from these tables that no apparent trend was visible in terms of realised prices by different size of farmers. In the case of paddy and maize the realised price was a little higher for the medium and large farmers in Punjab, while the advantage tilted towards marginal and small farmers in the case of cotton and potato in that state. Similarly in Bihar, in some crops, small or medium farmers obtained the highest price while in others, either marginal farmers or large farmers extracted the maximum price. In other words, there was no clear advantage in either class of size holdings.

Therefore, to draw a conclusion regarding farm size and realised prices, it becomes necessary to club the price of all the crops together to get some aggregate price index. It is not appropriate to take a simple average of the prices of different crops because price is already an average figure. The following procedure is adopted to club the prices of different crops:

$$P_j = \frac{\sum P_{ij} Q_{ij}}{\sum Q_{ij}}, \qquad i = 1, 2, \ldots, N \text{ (N crops)}$$

$$j = 1, 2, \ldots, M \text{ (M no. of households)},$$

Where

P = price obtained by the farmer,

Q = quantity sold by the farmer.

The price index of jth household is obtained by multiplying j[th] farmer's price obtained for each crop by the quantity sold of each crop and dividing the sum of all the crops by the sum of quantity sold of all crops. This is the indirect weighted price, the weight being the quantity sold. One can also apply direct weights to calculate the weighted price index. The weighted price index is calculated here by giving weight to the above price index in terms of area under a particular crop divided by the gross cropped area. The weighted price index is calculated as under:

$$P_j = \frac{\sum P_{ij} Q_{ij} W_{ij}}{\sum Q_{ij} W_{IJ}},$$

Where

W_{ij} represents weight for j[th] farmer for the ith crop. The results are given in Table 4.8.

Table 4.8

Aggregate Price Index (All Crops)

(Rs per quintal)

	Gross Price Index	*Weighted Gross Price Index*	*Net Price Index*	*Weighted Net Price Index*
Punjab				
Marginal	557	525	545	513
Small	501	493	491	484
Medium	546	591	540	586
Large	472	437	464	433
Total	486	443	478	439
Bihar				
Marginal	475	510	474	509
Small	346	385	342	382
Medium	262	279	254	271
Large	522	519	519	517
Total	367	432	362	427

Source: Computed using field survey data.

The results of the aggregate price index rule out any discrimination against the marginal or small farmers in both the states. It is clearly evident that in Punjab, marginal farmers obtained the highest price, gross as well as net, in terms of the unweighted price index and second highest in terms of the weighted price index. In Bihar, marginal farmers obtained the second highest price after large farmers, both in terms of gross and net and in terms of unweighted and weighted price indices. The lowest price was obtained by the large farmers, both in terms of gross and net and weighted and unweighted price indices in Punjab. In Bihar, the lowest gross and net price occurred in the case of medium farmers for weighted and unweighted price indices.

Thus, these findings have far reaching implications. In Punjab, most of the produce was sold through regulated *mandis*, thereby providing the farmers with a better marketing environment enabling them to realise better prices for their produce. Punjab farmers realised a competitive price in the market whereby the biggest beneficiary was the marginal farmer. Although the marginal farmer in Punjab was being squeezed by the rising cost of cultivation and stagnating yield, forcing him to quit farming by leasing out land in favour of large farmers, these findings clearly rule out any discrimination on account of the sale of his surplus produce in the market. Bihar on the other hand, lacked in basic market infrastructure in terms of regulated *mandis* and thereby 90 per cent of the output was being disposed off in the village to various intermediaries. Interlocking of credit and marketing of agricultural produce (Harris, 1991; Olsen, 1993; Selvaraj and Sundaravaradarajan, 1999; Bardhan, 1980) appear to be still operating. Because of lack of market infrastructure, there were hardly any government procurement operations taking place in the state thereby leaving the farmer at the mercy of private trade which was full of exploitation, with the end result that the realised prices were far lower than the minimum support prices. However, not only were the marginal and small farmers suffering because of lack of market reforms but the large farmers also lost in terms of lower realised prices for the surplus output produced by them.

Regression Results

The analysis presented in the previous section gave us an idea of the behaviour of marketed surplus, marketing channels followed, marketing costs and price received by the farmers for different crops across various farm size groups. In order to find out the determinants of marketed surplus and the price received by the farmers, a regression exercise was carried out.

Regression was done on the pooled data for all the categories rather than separate regressions for each farm size category. The results of the functional analysis are presented in the following sections.

Determinants of Marketed Surplus

Marketed surplus of any commodity at the farm level depends primarily on the amount of output produced by an individual farmer. Area operated or area under a particular crop could be used as a proxy for the output and therefore can be another determinant for marketed surplus. In addition to output and net operated area (NOA), the other independent variables in the study are: average household size (numbers), irrigated area as a proportion of gross cropped area, total loans per acre of area operated (Rs), area leased in as a proportion of net operated area, physical assets per acre (Rs) and a dummy variable for market channels with value one for those households who sold their output in the wholesale market and zero for others. In Bihar another dummy variable for the second channel is used, with a value of one for the households who sell their product in the nearby local market and zero otherwise. Log linear regressions have been used for the statistical estimation.

Analysing the results of aggregate output, it is observed from Table 4.9(a) and 4.9(b), respectively for Punjab and Bihar, that the value of output was the most important variable determining the value of marketed surplus. The coefficient of output was positive and significant at one per cent. The value of the coefficient was more than one in both the states, indicating that any rise in output led to a more than proportionate increase in marketed surplus. Household size was the other important variable significant at one per cent with a negative coefficient, indicating that an additional member in the family reduced marketed surplus, by raising the retention requirement by 16 per cent in Punjab and 25 per cent in Bihar. Replacing the value of output by the operated area, it is observed that operated area also has a significant positive relationship with marketed surplus. A one per cent increase in net operated area raised the marketed surplus by 1.3 per cent in Punjab and 1.1 per cent in Bihar. The variable for farm assets held by the households was significant in Punjab, indicating that a 10 per cent increase in their value (in the form of a tractor/other farm implements) led to rise in marketed surplus by 0.3 per cent. Area

Table 4.9a

Determinants of Marketed Surplus: Punjab

	Const	Log Output	Log Area	Log HH Size	Log Irrigation	Log Loan	Dummy-Channel	Log Assets	Log Leased-in	$\bar{R}^2$	F*
Aggregate	-2.48	1.20		-0.16	0.03	-	-	0.00	-0.04	0.92	698.7
	-(9.8)	(51.3)		-(3.2)	(0.7)			(0.4)	-(0.8)		
	9.26		1.28	-0.31	-0.10	-	-	0.03	-0.07	0.79	219.2
	(47.8)		(28.0)	-(3.7)	-(1.4)			(2.0)	-(0.9)		
Wheat	-2.05	1.43	-	-0.30	-0.01	0.02	0.41	-0.02	-	0.81	208.0
	-(7.9)	(22.4)		-(3.1)	-(1.2)	(2.2)	(4.5)	-(1.3)			
	1.69	-	1.40	-0.35	0.00	0.03	0.50	-0.01	-	0.75	146.3
	(6.8)		(17.7)	-(3.2)	-(0.5)	(2.1)	(4.8)	-(0.5)			
Paddy	-0.17	1.07	-	-0.05	0.00	0.00	-0.06	-0.01	-	0.93	508.4
	-(0.9)	(30.3)		-(1.1)	-(0.8)	(0.3)	-(0.6)	-(1.0)			
	3.25	-	1.06	-0.13	0.00	0.00	0.03	-0.01	-	0.89	298.9
	(18.2)		(22.0)	-(2.4)	(0.0)	(0.4)	(0.3)	-(0.6)			
Potato	0.11	0.96	-	0.02	0.00	0.01	0.02	-0.02	-	0.94	201.5
	(0.3)	(24.0)		(0.3)	-(0.5)	(1.2)	(0.2)	-(0.8)			
	4.77	-	1.03	-0.02	0.00	0.01	-0.12	-0.04	-	0.87	91.2
	(13.7)		(15.5)	-(0.3)	(0.7)	(1.2)	-(1.0)	-(1.2)			
Mustard	-1.51	0.78	-	0.11	0.06	0.03	0.56	0.05	-	0.75	13.2
	-(1.0)	(3.2)		(0.5)	(0.7)	(0.8)	(1.8)	(0.3)			
	0.15	-	0.68	0.01	0.12	0.03	0.69	0.01	-	0.69	9.8
	(0.1)		(2.1)	(0.0)	(1.4)	(0.6)	(2.0)	(0.1)			
Maize	-0.65	1.13	-	-0.48	-0.01	-	0.75	0.02	-	0.46	8.4
	-(0.8)	(3.3)		-(1.8)	-(0.1)		(2.0)	(0.3)			
	1.34	-	0.50	-0.48	0.17	-	0.76	0.07	-	0.34	5.5
	(1.8)		(1.5)	-(1.6)	(1.5)		(1.8)	(1.0)			
Cotton	-0.08	0.99	-	0.01	0.00	0.00	-	0.01	-	1.00	1848
	-(1.0)	(48.9)		(0.3)	-(0.2)	-(0.9)		(2.3)			
	1.98	-	1.15	-0.04	-0.01	0.00	-	-0.01	-	0.94	120.1
	(9.1)		(11.5)	-(0.4)	-(0.5)	-(0.4)		-(0.3)			

Note: 1) Figures in parentheses are respective t values.
2) The marked surplus and output are in value terms in rupees in the aggregate equations and in physical terms in quintals in crop-wise equations.

Source: Based on Field Survey data.

Table 4.9b

Determinants of Marketed Surplus: Bihar

Crop	*Const*	*Log Output*	*Log Area*	*Log HH Size*	*Log Irrigation*	*Log Loan*	*Log Assets*	*Log Leased-in Land*	*Dummy-Channel1*	*Dummy-Channel2*	$\bar{R}^2$	*F**
Aggregate	-2.73	1.20		-0.25	-0.16-	0.02	-	0.03	-	-	0.70	129.1
	-(5.7)	(24.5)		-(2.9)	(0.7)	(0.5)		(0.3)				
	8.39		1.10	-0.18	0.02	0.02	-	0.08	-	-	0.58	77.8
	(42.9)		(18.9)	-(1.9)	(0.1)	(0.6)		(0.6)				
Wheat	-1.17	1.14	-	-0.31	-0.47	0.08	-0.02	-	0.98	0.58	0.73	129.9
	-(6.6)	(22.8)		-(3.8)	-(0.5)	(2.3)	-(2.0)		(5.2)	(6.3)		
	1.32	-	1.05	-0.24	0.32	0.06	-0.03	-	1.11	0.70	0.65	87.1
	(6.8)		(17.7)	-(2.5)	(0.3)	(1.4)	-(2.5)		(5.1)	(6.6)		
Paddy	-0.81	0.28	-	0.08	-0.20	0.04	-	-	2.56	2.16	0.63	73.03
	-(3.8)	(5.7)		(0.8)	-(0.8)	(1.1)			(9.2)	(16.6)		
	-0.11	-	0.29	0.07	-0.12	0.04	-	-	2.60	2.15	0.62	72.7
	-(0.6)		(5.6)	(0.7)	-(0.5)	(1.2)			(9.4)	(16.4)		
Sugarcane	-0.39	0.99	-	0.01	-	-	0.01	-	0.38	0.35	0.99	397
	-(1.8)	(38.8)		(0.2)			(0.9)		(4.1)	(4.1)		
	3.78	-	0.94	0.15	-	-	0.04	-	0.81	1.18	0.73	11.64
	(3.9)		(6.4)	(0.5)			(0.9)		(1.7)	(2.8)		

contd...

...contd...

Crop	Const	Log Output	Log Area	Log HH Size	Log Irrigation	Log Loan	Log Assets	Log Leased-in Land	Dummy-Channel1	Dummy-Channel2	$\bar{R}^2$	F*
Potato	-0.89	0.64	-	-0.12	-	-	0.03	-	2.47	1.26	0.65	35.13
	-(2.1)	(8.0)		-(0.7)			(1.6)		(3.9)	(7.1)		
	0.31	-	0.01	0.01	-	-	0.03	-	4.34	1.45	0.39	12.94
	(0.6)		(0.2)	(0.0)			(0.9)		(5.6)	(6.3)		
Masoor	0.02	0.57	-	-0.10	-	-	-0.02	-	1.00	0.43	0.53	27.98
	(0.1)	(9.3)		-(1.1)			-(1.2)		(2.3)	(3.6)		
	0.42	-	0.40	-0.01	-	-	-0.01	-	1.32	0.54	0.32	12.45
	(1.5)		(5.0)	-(0.1)			-(0.7)		(2.5)	(3.8)		
Maize	-0.96	0.88	-	-0.11	-	-	-0.01	-	1.13	1.02	0.66	86.86
	-(3.3)	(14.8)		-(1.0)			-(0.3)		(4.7)	(7.1)		
	1.13	-	0.48	-0.20	-	-	-0.05	-	1.76	1.39	0.39	28.51
	(3.4)		(5.0)	-(1.3)			-(2.0)		(5.5)	(7.3)		
Khesari	-0.33	0.77	-	0.00	-	-	0.00	-	-0.70	0.25	0.61	28.43
	-(1.4)	(9.8)		(0.0)			-(0.2)		-(1.2)	(1.8)		
	0.54	-	0.66	0.02	-	-	-0.02	-	-0.48	0.49	0.47	16.14
	(1.8)		(6.9)	(0.1)			-(1.0)		-(0.7)	(3.2)		

Note: 1) Figures in parentheses are respective t values.

2) The marked surplus and output are in value terms in rupees in the aggregate equations and in physical terms in quintals in crop-wise equations.

Source: Based on Field Survey data.

irrigated, loans and area under tenancy were insignificant for the aggregate output in both the states.

As in the case of aggregate marketed surplus, output of each crop was also the most significant variable affecting the marketed surplus of the respective crops in both the states. The magnitude of the coefficient of output exceeded one in the case of the main crops of wheat and paddy in Punjab, indicating a more than proportionate rise in the marketed surplus with a rise in output of these crops. In the case of cotton, the third most significant crop in Punjab, the value of the coefficient was around unity, indicating a proportionate increase in the marketed surplus with any increase in the output of this crop. In Bihar on the other hand, except wheat, the coefficient of output was less than unity for all the major and minor crops. For the main staple crop of paddy, the coefficient was less than 0.3, indicating the subsistence nature of this crop. The other coarse grain and pulse crops with significant area, namely maize and *khesari*, were better marketed with the value of the coefficient ranging between 0.77 to 0.88.

In the alternate equations, the marketed surplus is regressed on the cultivated area of the crop instead of output. The results show that area was also highly significant for all the crops with a positive coefficient in both the states. The crop-wise area coefficient had a similar trend to that of output. The coefficient exceeded unit value for almost all crops in Punjab, but was less than unity in most of the cases in Bihar. The area results further corroborate that in Punjab, marketed surplus increases more than proportionately and in Bihar it increases less than proportionately with the rise in area. In the case of other variables, household size and marketed surplus had a significant negative relationship in the case of all major crops, except for some commercial crops like cotton, potato and sugarcane, in both the states. In Bihar however, intriguingly, household size was not significant in the case of paddy. The dummy variable for both market channels was significant in Bihar, indicating higher market surplus for those households who sell their produce in the regulated wholesale markets or in the local village market. In Punjab however, the dummy variable for market channel was not significant for most of the crops, as most of the farmers in Punjab sold their produce in the regulated markets. Finally, loans, assets and the variable for irrigation turned out to be insignificant for almost all the crops. Thus, the most important variables affecting the crop-wise marketed surplus were output and area cultivated of the respective crops, with a positive or direct effect, and family size with a negative or indirect effect. The dummy variable for the market channels was significant in Bihar.

Determinants of Prices Obtained by the Farmers

The results of the multiple regression analysis carried out to ascertain the volume and direction of factors affecting price formation are presented in Table 4.10(a) and 4.10(b). The price obtained by the farmer is taken as the dependent variable. The independent variables are: value of marketed surplus, (Rs), net operated area, (acres), value of output produced—excluding byproduct (Rs), storage quantity or period, loans (Rs), value of physical assets (Rs) and a dummy variable for marketing channels as defined in the earlier case described above. We analysed both crop marketed surplus, crop output and crop area as well as aggregate marketed surplus, aggregate output and aggregate operated area, and the results are presented with the best-fit regressions.

It is clear from the tables that the value of the marketed surplus and the value of the output, which also represent the economic status of the farmers, had significant coefficients for most of the crops regressed in both the states, with a positive sign, thereby indicating a rise in the net price received by the farmers with a rise in their economic status. The reason for a positive relationship seems to be that with a higher value of output and marketed surplus, the farmers were in a better position to choose the appropriate channel for marketing their produce (supported by the better transportation facilities).

The coefficient of storage was insignificant or even negative in a few cases, indicating that if farmers indulged in stocking it reduced the overall produce offered for marketing and thereby affected their price negatively. Similarly, the variable for assets was also insignificant in both the states. In Bihar, it could be because the farmers were generally selling their produce to the intermediaries within the village, at the local market or on the field itself. Therefore, price was independent of whether the farmers owned any particular transportation asset or not. In Punjab, on the other hand, because the majority of the farmers sold their produce in the regulated *mandis,* and because of the availability of metalled roads, owning a

Table 4.10(a)

Determinants of Price of Crops Received by the Farmers: Punjab

	Const	*Log Market Surplus*	*Log Output*	*Log Area*	*Log Loan*	*Log Assets*	*Dummy-Channel*	*Log Storage*	$\bar{R}^2$	*F**
Wheat	6.42 (675)	0.002 (2.5)	-	-	-0.000 -(0.4)	-0.000 -(0.2)	0.002 (0.4)	0.001 (0.6)	0.02	2.05
	6.37 (254)	-	0.008 (2.7)	-	-0.000 -(0.4)	-0.001 -(0.9)	0.004 (0.0)	0.001 (0.6)	0.02	2.24
	6.40 (560)	0.004 (4.0)#	-	-	-0.000 -(0.3)	-0.001 -(1.3)	0.003 (0.6)	0.001 (0.6)	0.05	4.05
	6.38 (252)	-	0.006 (2.4)#	-	-0.000 -(0.2)	-0.001 -(0.9)	0.005 (0.9)	0.002 (0.8)	0.02	1.93
Paddy	6.25 (125)	-	0.013 (2.5)	-	0.000 (0.5)	-0.002 -(1.4)	0.012 (0.6)	-0.004 -(2.0)	0.02	1.80
	6.36(248)	-	-	0.012 (2.3)	0.000 (0.5)	-0.002 -(1.4)	0.014 (0.7)	-0.004 -(1.9)	0.01	1.58
	6.26 (131)	0.01 (2.4)#	-	-	0.000 (0.5)	-0.002 -(1.4)	0.011 (0.6)	-0.004 -(2.0)	0.01	1.63
	6.22 (114.4)	-	0.01 (2.7)#	-	0.000 (0.4)	-0.003 -(1.6)	0.009 (0.5)	-0.005 -(2.2)	0.02	1.96
	6.36 (249)	-		0.01 (2.6)#	0.000 (0.3)	-0.003 -(1.6)	0.010) (0.5)	-0.004 -(2.0)	0.02	1.93

contd...

...contd...

	Const	Log Market Surplus	Log Output	Log Area	Log Loan	Log Assets	Dummy-Channel	Log Storage	$\bar{R}^2$	F^*
Potato	4.74	0.093	-	-	0.003	-0.025	-0.010	-0.003	0.04	1.57
	(9.6)	(2.5)			(0.4)	-(1.1)	-(0.1)	-(0.2)		
	4.65	-	0.101	-	0.004	-0.026	-0.003	-0.009	0.05	1.74
	(9.3)		(2.7)		(0.5)	-(1.1)	(0.0)	-(0.5)		
Mustard	7.12	0.005	-	-	0.002	0.015	-0.01	0.108	0.04	1.21
	(26.7)	(0.7)			(0.4)	(0.6)	-(0.2)	(1.6)		
	6.58	-	0.060	-	0.002	0.018	-0.056	0.095	0.29	2.99
	(22.3)		(2.7)		(0.4)	(0.9)	-(1.1)	(1.6)		
	6.66	-	0.049	-	0.001	0.005	0.019	0.087	0.16	1.88
	(19.1)		(1.8)#		(0.3)	(0.2)	(0.4)	(1.3)		
Maize	6.29	-	-	0.04	-0.001	-0.005	-0.006	0.005	0.08	1.76
	(169)			(2.4)#	-(0.5)	-(1.1)	-(0.2)	(0.8)		
	5.99	-	0.031	-	-0.002	-0.004	-0.005	0.004	0.05	1.51
	(44.5)		(2.2)#		-(0.6)	-(0.9)	-(0.2)	(0.6)		

Note: # denotes that in these equations, the aggregate variable is used while in all other equations the crop-level variable is used while fitting the regression equation.

Table 4.10(b)

Determinants of Price of Crops Received by the Farmers: Bihar

	Const	*Log market surplus*	*Log output*	*Log area*	*Log loan*	*Log assets*	*Dummy-channel1*	*Dummy-channel2*	*Log storage*	$\bar{R}^2$	*F**
Wheat	6.30 (128.0)	0.01 (1.7)	-	-	0.000 -(0.1)	0.001 (1.1)	0.01 (0.3)	0.01 (0.6)	-0.05 -(2.7)	0.02	1.76
	6.32 (149.8)	0.01 (1.7)#	-	-	0.000 (0.0)	0.001 (1.0)	0.02 (0.8)	0.01 (0.4)	-0.05 -(2.7)	0.02	1.75
Paddy	6.05 (507.4)	-	-	-0.01 -(2.1)	-0.02 -(3.4)	-0.002 -(1.6)	0.04 (1.1)	0.03 (1.9)	0.00 (0.0)	0.06	3.85
	6.13 (104.3)	-	-0.01 -(1.6)	-	-0.02 -(3.4)	0.00 (0.0)	0.04 (1.0)	0.03 (1.9)	0.00 -(0.2)	0.06	3.51
Potato	5.46 (27.2)	0.00 (0.0)	-	-	-	0.04 (2.0)	-0.27 -(0.4)	-0.11 -(0.4)	-	0.003	1.08
	3.16 (3.6)	-	0.30 (2.7)	-	-	0.03 (1.4)	-1.04 -(1.5)	-0.17 -(0.9)	-	0.08	3.00
	5.57 (25.1)	-	—	-0.06 -(1.0)	-	0.04 (2.0)	-0.33 -(0.5)	-0.13 -(0.6)	-	0.01	1.33
Masoor	7.34 (331.4)	0.01 (2.9)	-	-	-	-0.00 -(0.5)	0.00 (0.1)	-0.01 -(0.2)	-	0.10	4.21
	6.95 (70.6)	-	0.05 (4.4)	-	-	-0.00 -(0.7)	0.02 (0.3)	0.04 (1.7)	-	0.17	7.09
	7.37 (357.9)	-	-	0.02 (1.2)	-	0.00 -(0.8)	0.07 (0.8)	0.05 (2.2)	-	0.04	2.36
Maize	6.16 (377.0)	-0.01 -(2.5)	-	-	-	0.00 -(1.1)	0.03 (0.8)	0.07 (2.5)	-	0.02	2.11
Khesari	7.21 (38.2)	-	-0.03 -(1.9)#	-	-	0.00 (0.2)	-0.01 -(0.1)	-0.02 -(0.6)	-	0.00	1.04

Note: # denotes that in these equations, the aggregate variable is used while in all other equations the crop-level variable is used while fitting the regression equation.

transportation asset did not make much of a difference. In a few cases in Bihar, the loan variable was significant with a negative sign, indicating the inter-linkage between the credit and produce market affecting the farmers' price negatively. However, with the widespread network of regulated markets in Punjab, such interlinkages were missing completely in that state, so the coefficient for credit was insignificant for almost all crops in Punjab. Finally, the dummy variable for the market channel was insignificant in Punjab indicating no difference in price with respect to market channels, as more than 90 per cent of the output was marketed through regulated markets. In Bihar, the dummy variable for the first channel was insignificant indicating no price difference for those 10 per cent farmers who sold their product in the regulated *mandis*. That is probably the reason that farmers prefer to sell to the intermediaries. This is also reflected in the coefficient of the second dummy, which was significant in a few cases with a positive sign, indicating that farmers who sell in the village or in the local markets are getting higher prices than in the case of all other channels. Before concluding the discussion, it must be pointed out that the value of $\overline{R}^2$ was quite low in the price equations. The low value of $\overline{R}^2$ indicates that the explanatory variables are not able to explain much of the variation in the dependent variable and probably some important variables are not included in the model. It is well known that the price in the market is determined by a multiple number of factors like the number of traders present in the bidding, the total amount of output being auctioned, price expectations of the traders, size and location of the *mandi*, government price policy, etc. All these explanatory variables could not be included in the regression analysis. The regressions with low $\overline{R}^2$ are retained in the analysis because of their desired economic interpretation.

Annexure Table A-4.1

Marketed Surplus Per Household and Per Acre of Cropped Area in Punjab by Different Farm Size Categories

(Quintals per acre)

Crop	Category Operated	Output per Household	Output per Acre	Marketed Surplus per Household	Marketed Surplus per Acre	Marketed Surplus as a % of Output
Wheat	Marginal	26.5	14.5	13.0	7.1	49.2
	Small	52.9	14.0	35.5	9.4	67.2
	Medium	98.8	14.8	76.5	11.5	77.4
	Large	252.2	14.9	203.8	12.1	80.8
Mustard	Marginal	-	-	-	-	-
	Small	9.3	4.5	3.8	1.8	40.5
	Medium	6.8	3.9	5.8	3.3	85.2
	Large	27.7	6.4	19.9	4.6	72.0
Other vegetables	Marginal	175.0	87.5	175.0	87.5	100.0
	Small	21.3	25.9	21.3	25.9	100.0
	Medium	36.4	19.3	29.5	15.6	81.0
	Large	129.8	16.5	122.0	15.5	94.0
Paddy	Marginal	38.9	25.1	37.1	23.9	95.2
	Small	81.1	24.3	79.9	23.9	98.6
	Medium	142.0	24.4	135.4	23.3	95.4
	Large	444.1	24.9	436.6	24.5	98.3
Maize	Marginal	14.9	4.2	9.1	2.6	61.2
	Small	27.8	8.5	17.6	5.3	63.2
	Medium	35.5	7.1	19.5	3.9	54.9
	Large	49.9	8.4	37.4	6.3	75.0
Cotton	Marginal	13.0	7.4	11.0	6.3	84.6
	Small	29.8	9.9	21.3	7.1	71.4
	Medium	28.6	7.5	28.6	7.5	100.0
	Large	65.3	8.4	62.3	8.0	95.4

Source: Field Survey.

Annexure Table A-4.2

Marketed Surplus Per Household and Per Acre of Cropped Area in Bihar by Different Farm Size Categories

(Quintals per acre)

Crop	*Category Operated*	*Output per Household*	*Output per Acre*	*Marketed Surplus per Household*	*Marketed Surplus per Acre*	*Marketed Surplus as a % of Output*
Wheat	Marginal	28.1	10.2	9.0	3.3	31.9
	Small	27.3	9.7	12.5	4.4	45.7
	Medium	37.4	9.5	19.9	5.1	53.2
	Large	102.2	11.4	65.9	7.3	64.5
Sugarcane	Marginal	125.0	304.9	125.0	304.9	100.0
	Small	249.6	203.9	152.7	124.7	61.2
	Medium	491.0	208.1	488.7	207.1	99.5
	Large	-	-	-	-	-
Gram	Marginal	3.6	4.6	1.5	2.0	42.8
	Small	4.9	4.9	4.0	4.0	81.7
	Medium	7.3	4.3	4.2	2.5	57.3
	Large	6.4	1.7	2.7	0.7	41.9
Arhar	Marginal	6.0	2.1	2.0	0.7	33.6
	Small	12.3	8.0	5.0	3.3	40.8
	Medium	13.4	4.1	4.3	1.3	32.3
	Large	13.0	7.7	9.7	5.7	74.4
Khesari	Marginal	3.1	3.5	1.6	1.8	51.0
	Small	7.8	4.6	2.8	1.7	36.4
	Medium	10.6	3.8	7.9	2.9	74.3
	Large	13.2	2.8	9.1	1.9	69.4
Masoor	Marginal	5.0	4.0	2.8	2.3	56.9
	Small	5.5	3.2	2.6	1.5	46.7
	Medium	12.8	4.3	6.4	2.1	50.1
	Large	13.8	3.0	6.7	1.5	48.2
Mustard	Marginal	7.3	4.5	2.3	1.4	31.0
	Small	5.2	3.8	2.4	1.8	46.2
	Medium	10.6	3.0	2.0	0.6	18.9
	Large	10.4	4.0	2.4	0.9	23.3

contd...

...contd...

Crop	*Category Operated*	*Output per Household*	*Output per Acre*	*Marketed Surplus per Household*	*Marketed Surplus per Acre*	*Marketed Surplus as a % of Output*
Potato	Marginal	12.5	19.3	4.6	7.1	37.0
	Small	26.4	57.0	9.1	19.7	34.6
	Medium	42.2	82.9	26.8	52.7	63.5
	Large	59.8	59.0	43.0	42.4	71.9
Onion	Marginal	7.0	46.7	2.2	14.3	30.7
	Small	37.8	86.6	28.3	64.9	74.9
	Medium	43.0	46.7	38.2	41.5	88.9
	Large	52.5	44.5	44.0	37.3	83.8
Other *rabi* vegetables	Marginal	9.1	10.1	2.5	2.8	27.5
	Small	19.7	37.8	5.8	11.1	29.5
	Medium	3.0	20.0	0.0	0.0	0.0
	Large	25.0	10.2	18.0	7.4	72.0
Paddy	Marginal	77.1	12.7	8.6	1.4	11.2
	Small	122.4	9.4	14.3	1.1	11.6
	Medium	218.4	13.3	24.1	1.5	11.0
	Large	155.9	11.0	44.6	3.1	28.6
Other crop	Marginal	4.0	16.0	3.0	12.0	75.0
	Small	17.8	22.6	8.9	11.3	50.0
	Medium	43.0	31.4	28.0	20.4	65.1
	Large	120.0	17.9	0.0	0.0	0.0
Maize	Marginal	32.3	12.9	19.8	7.9	61.2
	Small	24.6	11.0	15.5	6.9	63.0
	Medium	55.3	10.8	27.1	5.3	49.0
	Large	106.1	12.3	65.1	7.6	61.4

Source: Field Survey.

Annexure Table A-4.3

Gross and Net Price Received and Marketing Cost of Major Crops in Punjab

(Rs per quintal)

Crop	*Category Operated*	*Gross Price per Quintal*	*Transportation Cost*	*Marketing Fee*	*Other Cost*	*Marketing Cost per Quintal*	*Net Price per Quintal*
Wheat	Marginal	624	7.98	1.03	0.31	9.32	615
	Small	627	5.77	0.63	0.15	6.54	620
	Medium	628	2.56	1.03	0.32	3.91	624
	Large	628	3.09	1.08	0.12	4.30	624
Mustard	Marginal	-	-	-	-	-	-
	Small	1553	26.67	0.00	0.00	26.67	1527
	Medium	1543	26.09	0.00	0.00	26.09	1517
	Large	1650	5.94	1.00	0.00	6.94	1643
Other *rabi* vegetables	Marginal	200	14.29	0.00	0.00	14.29	186
	Small	526	41.78	7.72	0.00	49.50	477
	Medium	765	10.84	0.75	0.00	11.59	753
	Large	374	8.52	2.66	0.00	11.18	363
Paddy	Marginal	572	8.30	1.47	2.34	12.11	559
	Small	563	4.96	0.45	2.13	7.53	556
	Medium	574	2.70	1.04	2.01	5.75	568
	Large	584	7.69	1.25	1.57	10.51	574
Maize	Marginal	534	9.02	0.00	3.90	12.93	521
	Small	537	5.21	0.00	1.04	6.26	530
	Medium	544	10.26	0.00	1.28	11.54	532
	Large	565	9.02	0.00	0.35	9.37	555
Cotton	Marginal	2445	22.73	0.00	5.91	28.64	2417
	Small	2365	72.94	0.00	0.00	72.94	2292
	Medium	2636	6.65	0.00	0.00	6.65	2629
	Large	2022	5.30	0.00	1.06	6.36	2015
Potato	Marginal	275	10.00	0.00	5.00	15.00	260
	Small	252	12.68	0.05	1.63	14.36	238
	Medium	240	8.51	1.08	1.27	10.87	229
	Large	241	2.62	0.05	8.46	11.13	229

Source: Field Survey.

Annexure Table A-4.4

Gross and Net Price Received and Marketing Cost of Major Crops in Bihar

(Rs per quintal)

Crop	*Category Operated*	*Gross Price per Quintal*	*Transportation Cost*	*Marketing Fee*	*Other Cost*	*Marketing Cost per Quintal*	*Net Price per Quintal*
Wheat	Marginal	601	0.00	0.00	0.00	0.00	601
	Small	596	1.54	0.16	0.13	1.83	595
	Medium	607	0.54	0.05	0.07	0.66	606
	Large	567	0.80	0.22	0.44	1.47	566
Sugarcane	Marginal	77	3.20	0.00	0.80	4.00	73
	Small	67	6.98	0.00	0.00	6.98	60
	Medium	69	9.04	0.00	0.07	9.11	60
	Large	-	-	-	-	-	-
Gram	Marginal	1757	0.00	0.00	0.00	0.00	1757
	Small	1773	0.00	0.00	0.00	0.00	1773
	Medium	1698	0.00	0.00	0.00	0.00	1698
	Large	1613	1.25	0.00	0.00	1.25	1611
Arhar	Marginal	1400	0.00	0.00	0.00	0.00	1400
	Small	1600	0.00	0.00	0.00	0.00	1600
	Medium	1454	0.00	0.00	0.00	0.00	1454
	Large	1569	0.00	0.00	0.00	0.00	1569
Khesari	Marginal	918	10.71	0.00	0.00	10.71	907
	Small	944	4.72	0.00	0.00	4.72	939
	Medium	925	0.00	0.00	0.00	0.00	925
	Large	932	6.85	0.00	0.00	6.85	925
Masoor	Marginal	1604	0.00	0.00	0.00	0.00	1604
	Small	1744	0.00	0.00	0.00	0.00	1744
	Medium	1704	0.00	0.00	0.00	0.00	1704
	Large	1611	0.08	0.00	0.00	0.08	1610
Mustard	Marginal	1432	0.00	0.00	0.00	0.00	1432
	Small	1508	0.00	0.00	0.00	0.00	1508
	Medium	1833	0.00	0.00	0.00	0.00	1833
	Large	1421	0.00	0.00	0.00	0.00	1421

contd...

...contd...

Crop	*Category Operated*	*Gross Price per Quintal*	*Transportation Cost*	*Marketing Fee*	*Other Cost*	*Marketing Cost per Quintal*	*Net Price per Quintal*
Potato	Marginal	342	0.00	0.00	0.00	0.00	342
	Small	350	0.00	0.00	3.28	3.28	346
	Medium	286	0.05	0.00	8.43	8.48	277
	Large	314	0.00	0.00	4.57	4.57	310
Onion	Marginal	239	2.33	3.49	0.00	5.81	233
	Small	484	1.03	0.00	0.00	1.03	483
	Medium	389	0.00	0.00	0.00	0.00	389
	Large	502	0.00	0.00	0.00	0.00	502
Other vegetables	Marginal	430	17.79	3.56	0.00	21.35	409
	Small	370	14.35	3.23	0.00	17.58	352
	Medium	-	-	-	-	-	-
	Large	350	0.00	0.00	0.00	0.00	350
Paddy	Marginal	464	1.33	0.00	3.65	4.98	459
	Small	423	0.88	0.00	0.00	0.88	422
	Medium	449	35.87	0.08	0.00	35.95	413
	Large	461	1.63	1.23	0.08	2.94	458
Other crop	Marginal	400	0.00	0.00	0.00	0.00	400
	Small	427	0.00	0.00	11.24	11.24	416
	Medium	400	0.00	0.00	0.00	0.00	400
	Large	-	-	-	-	-	-
Maize	Marginal	438	0.28	0.03	0.08	0.38	437
	Small	446	0.94	0.04	0.32	1.30	445
	Medium	464	0.34	0.00	1.05	1.40	463
	Large	423	2.01	0.22	1.22	3.44	420

Source: Field Survey.

5 Employment, Earnings and Consumption

Employment in agriculture is predominantly determined by the cropping pattern and the state of technology in the farming sector. Introduction of modern technology, especially in wheat and rice, in the late sixties and early seventies, unleashed the forces of change that influenced productivity and production and also had consequences on employment absorption. This had both positive and negative effects on employment. On the one hand, higher use of fertilisers, expansion in irrigated area and increased use of high yielding variety seeds gave rise to higher cropping intensity, thus raising land productivity and higher labour absorption (Grewal and Kahlon, 1974; Krishna, 1974; Raju, 1976; Bisliah, 1978). On the other hand, mechanisation of farm operations through tractors, threshers, combined harvesters etc., on a very large scale especially in the green belt of Punjab and Haryana led to a fall in labour intensity. Despite this fall in labour absorption, several researchers observed that total agricultural employment on mechanised farms has increased rather than decreased, primarily because of higher cropping intensity, labour intensive shift in crop-mix and higher productivity, all factors adding to higher labour use (Sidhu and Grewal, 1990; Parihar and Sidhu, 1986; Oberai and Ahmed, 1981).

In the present chapter, a comparative analysis of farm and non-farm employment is undertaken in the states of Punjab and Bihar. Total earnings from farming and non-farming activities are estimated. Finally, to bring out the differences in the states, consumption pattern of food, non-food and other non-durable commodities is presented. The chapter is divided into three sections, one section being devoted to each of these themes. Section I begins with a discussion on the employment aspects of the selected households. The employment status and total employment in farming and non-farming activities are the issues discussed in this section. Section II presents the total earnings from farm and non-farm employment, including the earnings from casual labour. Section III deals with the differences in the consumption behaviour of the selected households.

Employment Status of Selected Households

It was observed in Chapter 3 that on an average only one-third of the members of the family were working and the remaining two third were either below the working age, unemployed, retired or disabled persons not fit for any economic activity and are only doing household activities. Out of this 30 to 35 per cent working population, around 23 per cent in Punjab and 21 per cent in Bihar were employed in agriculture and agriculture related activities (Table 5.1). In Punjab the percentage was high with 32 per cent for medium farmers and low, around 7 per cent, for landless labourers. In Bihar, the percentage was highest among the small farmers group—26 per cent—and lowest—11 per cent—for the landless labourers.

Table 5.1

Employment Status in the Two States

(Per cent)

	Percentage of Members in Non-Working Age	*Self-Employed in Agri. and Other Activities*	*Salaried*	*Ad hoc and Casual Labour*
Punjab				
Marginal	67.6	24.2	2.5	5.7
Small	69.3	25.7	3.0	2.0
Medium	66.4	31.7	1.9	0.0
Large	68.7	29.1	2.2	0.0
Landless	67.8	7.0	7.8	17.4
Total	68.1	22.6	3.8	5.6
Bihar				
Marginal	64.6	21.4	7.2	6.9
Small	68.4	25.7	4.6	1.2
Medium	70.6	21.4	7.2	0.9
Large	72.0	22.8	5.2	0.0
Landless	65.0	10.5	5.4	19.2
Total	67.4	21.1	6.2	5.3

Source: Field Survey.

In non-farm employment, around 4 per cent in Punjab and 6 per cent in Bihar were employed in regular government, semi-government or private sector salaried jobs. The percentage of such people was highest among the landless labourers in Punjab and marginal and medium farmers in Bihar. Predictably, landless labourers had the maximum percentage of members seeking their daily earnings as ad hoc and casual labourers in both the states. Their percentage—of casual labour—was 17 and 19, respectively in Punjab and Bihar. Thus, these figures on employment reveal the fact that whereas maximum number of family members of large and medium farmers were absorbed in agriculture and allied activities, a comparatively higher number of family members in the case of marginal and landless labourers tried to find employment in allied and non-farm activities, such as ad hoc and casual labourers and as regular salaried employment where the earnings were much higher.

Employment in Crop Sector

Labour absorption capacity of agriculture is on the decline in Punjab as well as in other Green Revolution areas where rapid mechanisation has taken place. Although employment intensity in the case of individual crops started declining since the early seventies, the overall labour absorption increased during that period on account of increasing gross cropped area and shift in cropping pattern from low labour absorption crops towards high labour absorption crops, e.g. a massive shift from coarse grains to paddy in Punjab and Haryana. The trend of overall increase in labour absorption in the 1970s was reversed in the 1980s and 1990s. The major cause of this decline was the sharp decrease in man-days required in the major crops like paddy, wheat, rapeseed, mustard, maize, *bajra*, sugarcane and *jowar* (Gill, 2002). As cropping intensity increased at a slow pace in the post eighties and cropping pattern almost stabilised with paddy-wheat rotation, labour absorption started declining during eighties and nineties.

Table 5.2 presents family and hired labour used in sowing, transplanting/caring and harvesting activities for the total cropped area for the selected households in Punjab and Bihar. It is seen from the table that harvesting activities involved the highest man-days, followed by caring, sowing and marketing activities, in that order. The total labour absorption in harvesting, sowing and other activities in Punjab were calculated as 40 man-days per acre. The number of man-days per acre declined from 50 in the case of marginal farmers to 43 on small farms and further to 35 in the case of medium farmers and again increased to 41 in the case of large farmers. In Bihar, on an average 43 man-days per acre were employed in the harvesting and sowing activities and there was a clear negative relation between labour employment and increasing farm size. The labour absorption was above 65 man-days per acre on marginal farms, which declined to 46 man-days on small farms, 42 man-days on medium farms and further to 32 man-days in the case of large farms. Thus, labour absorption was much higher in Bihar compared to Punjab up to medium farm size and was less than Punjab in the case of large farms. The explanation for the above trend lies in the fact that mechanisation in Punjab has occurred almost equally for all size classes. In Bihar on the other hand, mechanisation is completely lacking in the lower strata of the farming community and agriculture in their case is almost manual. On large size farms in Bihar, cropping intensity is low and farmers mostly depend on either casual labour for harvesting and sowing operations, or use a higher number of tractors and tractor operated implements, as participation of family labour in their case is very low.

Use of family labour shows an inverse relationship with farm size in both the states. In Punjab, around 33 man-days of family labour per acre were engaged in crop production in the case of marginal farmers, which decreased to a mere 5 man-days in the case of large farmers. Similarly in Bihar, 45 man-days of family labour per acre were used in crop production among marginal farmers, but it was as low as 9 man-days in the case of large farmers. There could be two reasons for the occurrence of this inverse relationship between family labour employment and farm size. The first is that absolute area increases with the increase in farm size, consequently showing lower family employment per acre.[1] The second reason could be that in the upper stratum, farmers are economically well-off and their offspring being better educated, look for employment opportunities in white collar jobs, self-employment in other activities and other non-farm employment. Consequently intensity of family labour in crop production declines in the case of medium and large size farmers.

1. Although with rise in farm size, family size also increases, rise in area is proportionately much higher than increase in family size. Consequently, family labour employed per acre reverses in the upper echelon of farm size.

Table 5.2

Use of Human Labour in Agricultural Activities

(Man-days per acre)

	Marginal	*Small*	*Medium*	*Large*	*Total*
Punjab					
		Family Labour			
Sowing	3.17	2.66	2.15	1.33	1.69
Caring	11.62	8.84	6.61	2.46	4.24
Harvesting/threshing	17.23	8.57	4.58	1.34	3.37
Marketing	1.15	0.87	0.59	0.36	0.49
Total	33.17	20.95	13.93	5.49	9.80
		Attached Labour			
Sowing	0.06	0.13	0.46	1.36	1.02
Caring	0.30	0.71	1.61	4.76	3.61
Harvesting/threshing	0.33	0.42	0.64	2.18	1.65
Marketing	0.00	0.02	0.06	0.20	0.15
Total	0.69	1.28	2.78	8.50	6.43
		Hired Labour			
Sowing	5.39	7.42	7.44	7.82	7.61
Caring	3.00	4.25	3.65	4.52	4.29
Harvesting/threshing	6.59	9.33	7.41	13.58	11.82
Marketing	0.63	0.16	0.26	0.64	0.53
Total	15.60	21.16	18.75	26.56	24.24
Bihar					
		Family Labour			
Sowing	11.19	5.90	4.51	2.37	5.08
Caring	14.12	7.60	4.59	2.87	6.07
Harvesting/threshing	19.09	9.90	6.56	3.13	7.98
Marketing	1.06	0.63	0.37	0.22	0.48
Total	45.46	24.03	16.03	8.59	19.61
		Attached Labour			
Sowing	0.01	0.07	0.14	0.40	0.19
Caring	0.00	0.07	0.09	0.26	0.13
Harvesting/threshing	0.10	0.14	0.17	0.40	0.23
Marketing	0.00	0.00	0.03	0.03	0.02
Total	0.12	0.28	0.43	1.09	0.57
		Hired Labour			
Sowing	5.82	6.78	8.12	5.44	6.63
Caring	2.73	3.95	3.62	3.86	3.64
Harvesting/threshing	11.25	11.04	13.13	12.60	12.22
Marketing	0.17	0.38	0.53	0.25	0.36
Total	19.97	22.14	25.39	22.15	22.85

Source: Field Survey.

The ratio of hired labour to family labour was much higher in Punjab as compared to Bihar. Out of total man-days employed in crop production in Punjab, only 24 per cent were family labour, and the remaining 76 per cent were hired labour. Out of total hired labour, 16 per cent came through attached labour for the whole year and the remaining 60 per cent were temporarily hired at the time of sowing, caring and harvesting. In Bihar, the contribution of family labour was much higher—around 46 per cent—and the remaining 54 per cent came through hired labour. Unlike Punjab where farmers had a tendency to hired labour on a regular basis for the whole year, in Bihar the proportion of attached labour was only one per cent and the remaining hired labour was on a temporary basis.

Employment in Livestock Activity

Among the allied agriculture activities, dairy farming or livestock rearing is the prominent one. Among our sample households, livestock rearing was the major source of employment as is indicated by the total man-days employed in livestock in Table 5.3. There were three main functions on which labour man-days were concentrated in animal rearing namely, collection and cutting of fodder and grazing the animals, looking after animals including the milching activities, and collection and disposing off of cow dung. Whereas in Punjab, all these activities absorbed an almost equal proportion of labour; in Bihar the labour used for collection of fodder exceeds that of the other two activities. The principal reason seems to be that Bihar farmers generally transported fodder as head-loads due to lack of proper road and transportation systems. This involved much higher labour time compared to Punjab, where farmers generally owned bullock carts or tractor trolleys accompanied by a good road setup.

In order to make a comparison of labour absorption in crop and livestock sectors, distribution of man-days per household for both these activities is presented in Annexure Table A-5.1 and the percentage distribution in Annexure Table A-5.2. It is evident from these tables that the manpower employed in the livestock sector comes mainly from family labour. In Punjab, family labour contributed about 70 per cent time to livestock activities and only 30 per cent was devoted to crop sector activities. Similarly, in Bihar the ratio of livestock and crop activities was

Table 5.3

Use of Human Labour in Livestock Rearing

(Man-days per household)

	Marginal	*Small*	*Medium*	*Large*	*Landless*	*Total*
Punjab						
	Collection and Cutting of Fodder					
Family labour	53.64	60.00	52.44	50.64	52.56	53.76
Permanent labour	2.76	10.44	23.16	46.32	1.80	15.48
Hired labour	0.36	0.48	0.00	1.68	0.00	0.48
Total	56.76	70.8	75.6	98.64	54.36	69.72
	Looking after Animals					
Family labour	52.08	63.96	70.08	76.92	39.60	58.20
Permanent labour	1.56	7.44	19.68	39.00	1.44	12.60
Hired labour	0.48	0.24	0.00	1.80	0.00	0.48
Total	54.12	71.64	89.76	117.72	41.04	71.16
	Collection of Cow Dung and making Dung Cakes					
Family labour	37.80	45.12	44.28	34.32	28.92	36.84
Permanent labour	0.72	3.48	7.92	24.48	0.00	6.84
Hired labour	0.84	0.84	6.24	11.52	0.00	3.48
Total	39.48	49.56	58.56	70.32	28.92	47.04
Bihar						
	Collection and Cutting of Fodder					
Family Labour	124.98	41.81	42.82	36.17	10.93	69.97
Permanent Labour	0.51	0.28	0.00	14.14	0.00	1.22
Hired Labour	0.00	0.63	0.89	2.07	0.46	0.48
Total	125.49	42.72	43.71	52.38	11.39	71.66
	Looking after Animals					
Family Labour	28.30	37.25	39.39	30.52	9.66	28.58
Permanent Labour	0.47	0.28	0.00	7.93	0.00	0.78
Hired Labour	0.00	0.00	0.00	1.55	0.39	0.17
Total	28.77	37.53	39.39	40.00	10.05	29.54
	Collection of Cow Dung and making Dung Cakes					
Family Labour	16.03	21.35	22.24	14.31	5.43	16.03
Permanent Labour	0.27	0.28	0.00	5.69	0.00	0.55
Hired Labour	0.06	0.28	0.81	4.72	0.66	0.63
Total	16.35	21.92	23.05	24.72	6.09	17.21
	Others					
Family Labour	6.23	1.33	1.00	2.41	1.51	3.38
Permanent Labour	0.00	0.34	0.00	1.03	0.00	0.14
Hired Labour	0.00	0.00	0.00	0.00	0.00	0.00
Total	6.23	1.67	1.00	3.45	1.51	3.52

Source: Field Survey.

64 and 36 per cent, respectively. Hardly any labour was hired on a temporary basis for livestock rearing as 97 per cent of the time of hired labour in Punjab and 98 per cent of their time in Bihar was spent on crop sector activities alone. In the case of attached labour, almost equal proportion of their time was devoted to both activities in Punjab, while in Bihar hardly any attached labour was used for either of the activities. Thus, it can be concluded from the given statistics that livestock rearing was the second important activity among the cultivators, which provided a major source of employment to the family labour in both the states. Comparing manpower in crop and livestock activities in these two states, it is observed from the data presented in the Annexure tables that in Punjab, around 59 per cent manpower was engaged in the crop sector and the remaining 41 per cent found employment in the livestock sector (Table A-5.2). In Bihar the ratio of crop and livestock employment was 54 and 46, respectively.

Comparing within the farm size categories, it is observed from Annexure Table A-5.2 that small and marginal farmers devoted comparatively more time to livestock rearing, while large farmers allocated more time to the management of crop production. A review of the percentage time period allocated to livestock activities clearly indicates a declining trend of time allocated to livestock across farm size in both the states. It is observed from the table that marginal farmers allocated 66 per cent of total labour time to livestock activities while large farmers allocated only 24 per cent of labour time to livestock in Punjab. In Bihar too, a similar situation was observed where marginal farmers devoted around 70 per cent of the man-days to livestock while large farmers had only 19 per cent labour time allocated to the livestock activities. The implication is that marginal and small farmers having small pieces of land have more spare time which they devote to livestock through which they try to supplement their farming income (as will be seen in the subsequent section). Large farmers' priority remains the crop sector, as the latter is also their main source of earning.

Farm and Non-Farm Earnings

Apart from farm business income, the other main sources of earnings of the selected households were dairy and poultry farming (livestock rearing), self-employment, non-farm employment, casual labour, and remittances obtained from outside. The main components of farm business income were discussed in Chapter 3. This section presents a detailed analysis of the other sources of earnings. The end of this section presents a comparative picture of the contribution of each source to the total household income.

Table 5.4 presents net income from livestock activities. Net income was calculated after subtracting material cost from the gross earnings from milk and other milk and poultry products. It is seen from the table that all sizes of farmers try to supplement their farming income with livestock proceeds. However, it is essential to clarify here that total earnings in the present case also includes the imputed value of milk and other products which are consumed at home and are not sold in the market. Further, the cost of items includes only the material cost of fodder etc., and no imputed value of labour charges is added to the total cost, as most of livestock activities are performed by family labour (see the previous section).

Bihar lagged far behind Punjab in earnings through livestock activity, as was also the case with farm business income. The average net household income from livestock activities was less than one-fourth in Bihar as compared to Punjab. On an average, each household was earning Rs 3703 in Bihar, compared to Rs 17,229 in Punjab, from dairy farming and related activities. It was pointed out in the previous section that small and marginal farmers devoted more time to livestock activities than large farmers. The outcome of higher time allocation is realised in terms of higher relative earnings from livestock activities in the case of small and marginal farmers. Per household farming income in the lower stratum too was low compared to large farmers. However, the difference in per household earnings from livestock activities between small and large farmer was not that high in both the states. Thus, one can conclude that smaller size farmers supplemented their farming income through livestock activities more than that of larger size farmers, in both the states.

Table 5.4

Net Income from Livestock Activities

(Rs per household)

	Marginal	*Small*	*Medium*	*Large*	*Landless*	*Total*
Punjab						
			Cost Items			
Green fodder (home)	5493	9520	6873	17894	2338	8012
Green fodder (purchased)	565	1188	1762	1496	1961	1440
Dry fodder (home)	3138	6838	8158	22484	2745	8279
Dry fodder (purchased)	1198	2481	3354	2745	2686	2480
Concentrates (purchased)	151	255	275	1422	50	412
Medicines	895	1501	1263	2975	461	1350
Total	11440	21783	21686	49016	10241	21974
			Value of Output (Sold/Home Consumed)			
Milk	23021	36591	33949	75294	20137	36712
Ghee	429	686	510	3214	684	1119
Cheese	5	10	17	645	16	138
Cow dung	708	1286	1049	2543	689	1226
Egg/poultry	0	4	5	12	10	7
Goat	0	0	0	0	2	1
Total	24163	38578	35531	81709	21538	39204
Net income	12722	16795	13845	32693	11297	17229
Bihar						
			Cost Items			
Green fodder (home)	934	1205	2068	2069	284	1114
Green fodder (purchased)	6	0	27	0	0	6
Dry fodder (home)	2567	3547	4101	5144	298	2762
Dry fodder (purchased)	133	262	701	0	230	249
Concentrates (home)	579	1077	1652	2791	183	915
Concentrates (purchased)	665	825	820	1855	356	746
Other fodder (purchased)	11	8	25	52	16	16
Medicines	90	149	414	195	48	148
Other expenditure	20	15	21	53	10	20
Total	5004	7087	9829	12159	1424	5975
			Value of Output (Sold/Home Consumed)			
Milk	7143	10141	15765	20934	2849	9171
Ghee	16	61	51	241	0	42
Cheese	0	0	0	0	0	0
Cow dung	339	377	683	1056	105	403
Poultry	1	0	0	0	0	0
Goat	5	0	0	0	0	2
Others	0	0	0	103	295	59
Total	7504	10579	16499	22335	3249	9678
Net income	2500	3492	6670	10176	1825	3703

Source: Field Survey.

Table 5.5 and 5.6 show the household income from self-employment and non-farm employment. It is clear from the tables that all categories of farmers in Bihar had higher net earnings under these two heads compared to Punjab farmers. Earnings from self-employment were from small businesses within the village, or in the surrounding villages or small towns/cities, e.g., village shops, hullers or shellers, commission agents (*arhatias*) shops etc. Under non-farm employment, the major activities were government or semi-government regular jobs, or jobs with any private company, small businesses, small industry etc. It was seen in the previous analysis that in Bihar farm productivity/income was less than one-fifth and livestock income was less than one-fourth of Punjab. Farmers in Bihar seek other means to supplement their agriculture income in non-agricultural activities. Educated youths look for white-collar jobs. However, opportunities being limited, they are forced to opt for self-employment in shops, businesses or other small industries. In Punjab, agriculture and allied activities fetch comparatively higher earnings and thus the percentage of people moving out of agriculture to other means of self-employment means is less than in Bihar.

Table 5.5

Self-Employment Earnings during the Current Year

(Per household)

	Marginal	*Small*	*Medium*	*Large*	*Landless*	*Total*
Punjab						
Family labour (man-days)	49	22	14	25	26	27
Net income (Rs)	6109	1898	1379	4212	3735	3585
Bihar						
Family labour (man-days)	48	99	95	125	40	69
Net income (Rs)	5847	15794	24621	31086	5158	12170

Source: Field Survey.

Table 5.6

Income from Regular Non-Farm Employment

(Rs per household)

	Marginal	*Small*	*Medium*	*Large*	*Landless*	*Total*
Punjab						
Govt./semi-govt. job	4726	8950	13200	12438	261	6932
Large private company	1979	2195	890	2077	290	1400
Small business/industry	312	1049	3931	0	97	822
Others	1278	1046	1448	706	8677	3296
Total	8295	13240	19469	15220	9326	12450
Bihar						
Govt./semi-govt. job	3368	10486	27002	26500	6102	10276
Large private company	5692	4665	10408	37362	3197	7857
Small business/industry	4187	3477	2187	0	211	2768
Family business/industry	507	0	1490	2069	0	561
Others	1459	1364	645	0	276	1015
Total	15214	19993	41732	65931	9787	22477

Source: Field Survey.

Table 5.7 shows earnings from the casual labour. The main source of full time earning for landless cultivators and labourers was casual labour. Besides landless labourers, marginal and small farmers (and a few medium farmers in the case of Bihar) also engaged in wage earnings during the slack season, to supplement their farm income. The first column of the table presents the different occupations defined in the National Industrial Classification (NIC) and the second column specifies the related activities as defined in the National Classification of Occupations (NCO). It is observed from the table that although there were around eight different agricultural and non-agricultural activities engaging casual labour, the major concentration remained on agricultural labourers in the agricultural sector, and on construction in the non-agriculture sector. Out of the total employment as casual labour in Punjab, wage earning in agriculture remained the principal activity, which provided around 85 per cent of total earnings in this category. Construction in the rural and surrounding urban areas provided 11 per cent of the earnings to the casual labourers in

Punjab. In Bihar, these two activities remained the principal source of employment. Though construction had a comparatively higher ratio of earnings of around 30 per cent while the share of agricultural labour was 63 per cent. Besides these two major activities, the other minor activities sharing the remaining 4 to 6 per cent of the casual earnings were transport equipment operations, house keeping, carpentry, wood preparation, hairdressing, etc.

In addition to non-farm employment and casual labour, the sample households also obtained some remittances from the family members working away from the village, either in neighbouring town/city or other states, or even abroad in the case of Punjab. Annexure Table A-5.3 gives the details of the household members who changed their occupation and shifted to other villages, cities or abroad and the reasons thereof. The incidence of change in occupation and migration was prevalent on a much higher scale in Bihar as compared to Punjab. Out of total family members in Punjab, only 1.9 per cent shifted their occupation or place of work, while in Bihar the migration was 8.8 per cent. A majority of migration in Punjab was among the literate youth who went abroad for daily wage earning, either to the Middle East, Europe, Canada or the United States. A small percentage of marginal farmers and landless labourers also migrated to the neighbouring villages in search of daily earnings. In the case of Bihar, migration was mostly to the urban areas in the neighbouring states. The reason for migration in both the states in majority of the cases was more earnings, while job transfers were also a reason quoted by a number of households in Bihar. Migration was either temporary or circular in two-thirds of the cases and only one-third changed their place/occupation permanently. In case of migration to other countries in Punjab, the migrated persons generally worked abroad for a designated time period and then returned to their family work. In Bihar, generally working persons in the family would migrate to other cities or states, e.g., Delhi, Mumbai, Calcutta, Punjab, Haryana, for some time period, and send their saved money to the non-working members who stayed back in Bihar.

It is seen from Table 5.8 that outside remittances per household were evenly distributed among all size of holdings in both the states, indicating that family members of all size of farmers had the tendency to look for

Table 5.7

Earnings from Casual Labour in Different Occupations during the Reference Year

(Rs per household)

National Industrial Classification (NIC) – 1998	*National Classification of Occupations (NCO)–1968*	*Marginal*	*Small*	*Medium*	*Landless*	*Total*
Punjab						
Agriculture, hunting and related service activities	Agricultural labourers	3013 (72.2)	1408 (76.3)	0 (0.0)	9975 (88.7)	5611 (85.1)
Forestry, logging and related service activities	Forestry workers	0 (0.0)	0 (0.0)	0 (0.0)	38 (0.3)	17 (0.3)
Manufacture of wood and wood products	Wood preparation workers and paper makers	0 (0.0)	0 (0.0)	0 (0.0)	92 (0.8)	41 (0.6)
	Carpenters, cabinet and related wood workers	232 (5.6)	0 (0.0)	0 (0.0)	0 (0.0)	62 (0.9)
Construction	Bricklayers and other construction workers	927 (22.2)	356 (19.3)	0 (0.0)	868 (7.7)	724 (11.0)
Land transport, transport via pipelines	Transport equipment operators	0 (0.0)	82 (4.4)	0 (0.0)	0 (0.0)	21 (0.3)
Private household with employed person	Maids and other housekeepers service workers	0 (0.0)	0 (0.0)	0 (0.0)	177 (1.6)	79 (1.2)
	Building caretakers, sweepers, cleaners and related workers	0 (0.0)	0 (0.0)	0 (0.0)	92 (0.8)	41 (0.6)
Total		4171 (100.0)	1846 (100.0)	0 (100.0)	11243 (100.0)	6596 (100.0)

contd...

...contd...

Bihar						
Agriculture, hunting and related service activities	Agricultural labourers	4033 (63.3)	707 (41.6)	230 (41.7)	7319 (64.7)	3386 (62.9)
Manufacture of wood and wood products	Wood preparation workers and paper makers	62 (1.0)	0 (0.0)	0 (0.0)	57 (0.5)	38 (0.7)
	Carpenters, cabinet and related wood workers	0 (0.0)	202 (11.9)	0 (0.0)	226 (2.0)	76 (1.4)
Construction	Bricklayers and other construction workers	2040 (32.0)	791 (46.5)	322 (58.3)	2943 (26.0)	1616 (30.0)
Land transport, transport via pipelines	Transport equipment operators	128 (2.0)	0 (0.0)	0 (0.0)	434 (3.8)	149 (2.8)
Other business activities	Labourers, n.e.c.	5(0.1)	0(0.0)	0(0.0)	132(1.2)	31(0.6)
Recreational, cultural and sporting activities	Broadcasting and sound equipment operators and cinema projections	41 (0.6)	0 (0.0)	0 (0.0)	0 (0.0)	17 (0.3)
Other service activities	Hair dressers, barbers and related workers	67 (1.0)	0 (0.0)	0 (0.0)	207 (1.8)	73 (1.4)
Total		6376 (100.0)	1700 (100.0)	552 (100.0)	11318 (100.0)	5386 (100.0)

Note: Figures in parentheses are respective percentage of total casual earnings; n.e.c. - not elsewhere classified.

Source: NIC Classification and Field Survey Data.

better earnings outside their village within the state or in other states or abroad. As going abroad requires huge expenditure on the intermediaries involved in the process, landless labourers due to lack of funds had the minimum incidence of migration abroad and corresponding nominal remittances in Punjab. The average amount of remittances was higher in Bihar as their migration percentage was also much higher compared to Punjab. Remittances received by all size classes were generally spent on consumer durables or non-durables. The productive use of such remittances for either buying land or purchasing other productive assets was minuscule.

Table 5.8

Outside Remittances Obtained and its Uses by the Selected Households

Category	*Remittance Rs per Household*	*Uses of Remittances (% of Households)*						
		Debt Payment	*Medical Treatment*	*Purchase of Land*	*Purchase of Consumer Non-Durable*	*Consumer Durable*	*Other Productive Assets*	*Others*
Punjab								
Marginal	2208	10.0	0.0	0.0	80.0	10.0	0.0	0.0
Small	2198	20.0	0.0	0.0	60.0	20.0	0.0	0.0
Medium	5172	0.0	0.0	0.0	100.0	0.0	0.0	0.0
Large	3529	0.0	0.0	0.0	100.0	0.0	0.0	0.0
Landless	19	0.0	0.0	0.0	100.0	0.0	0.0	0.0
Total	2236	10.0	0.0	0.0	80.0	10.0	0.0	0.0
Bihar								
Marginal	4603	9.7	24.2	3.2	40.3	21.0	1.6	0.0
Small	1773	4.8	4.8	0.0	42.9	38.1	9.5	0.0
Medium	11600	4.8	4.8	0.0	47.6	23.8	4.8	14.3
Large	1862	0.0	0.0	0.0	50.0	33.3	16.7	0.0
Landless	1404	31.3	6.3	0.0	43.8	12.5	6.3	0.0
Total	4282	10.3	14.3	1.6	42.9	23.8	4.8	2.4

Source: Field Survey.

Table 5.9(a) and 5.9(b) summarise the above discussion on farm and non-farm earnings by the selected households in both the states. Looking at the results for Punjab first, it is seen from the data that farm income was the chief source of earning in Punjab, which contributed around 76 per cent of total income for the selected households. In addition to farm income, the other agricultural allied activities, namely livestock rearing also contributed around 9 per cent of the earnings, thus making the share of agriculture and allied activities 85 per cent. Among the non-agricultural employment, government/semi-government jobs, casual labour, self-employment and remittances from outside were the other major source of household earnings.

Table 5.9(a)

Percentage Distribution of Income of the Households by Activities: Punjab

(in Rs)

	Marginal	*Small*	*Medium*	*Large*	*Landless*	*Total*
1. Farm business income	33.5	53.8	69.4	86.2	0.0	76.2
2. Net income from livestock	22.1	17.6	9.3	7.0	29.9	8.6
3. Earnings from agriculture related and other activities	12.1	10.7	7.2	2.7	5.7	3.9
3.1 Rent (leasing out land)	2.2	1.4	0.2	0.0	4.5	0.5
3.2 Rent (leasing out implement)	0.2	0.0	0.0	0.0	0.0	0.0
3.3 Interest on lending	0.0	0.9	0.0	0.6	0.0	0.4
3.4 Pension (old age + others)	5.8	4.8	3.5	0.4	0.4	1.3
3.5 Remittances from outside	3.8	2.3	3.5	0.8	0.1	1.1
3.6 Other receipts	0.0	1.3	0.0	0.9	0.7	0.6
4.0 Non-farm employment earnings	14.4	13.9	13.1	3.2	24.7	6.2
4.1 Govt./semi-govt. job	8.2	9.4	8.9	2.7	0.7	3.5
4.2 Large private company	3.4	2.3	0.6	0.4	0.8	0.7
4.3 Small business/industry	0.5	1.1	2.6	0.0	0.3	0.4
4.4 Others	2.2	1.1	1.0	0.2	23.0	1.6
5. Self-employment earnings	10.6	2.0	0.9	0.9	9.9	1.8
6. Earnings from casual labour	7.2	1.9	0.0	0.0	29.8	3.3
7. Gross total per household	100.0	100.0	100.0	100.0	100.0	100.0
8. Gross total per capita per annum	11055	14535	22963	55951	7010	31637

Note: Figures are respective percentages of gross household income.

Source: Field Survey.

Table 5.9(b)

Percentage Distribution of Income of Households by Activities: Bihar

(in Rs)

	Marginal	*Small*	*Medium*	*Large*	*Landless*	*Total*
1. Farm business income	21.7	38.3	33.7	46.6	0.0	36.1
2. Net income from livestock	4.8	4.8	3.9	3.9	5.6	4.1
3. Earnings from agriculture related and other activities	20.6	5.9	22.9	12.5	14.4	15.4
3.1 Rent (leasing out land)	0.6	0.4	0.2	1.0	5.1	0.8
3.2 Rent (leasing out) implement	1.5	0.5	1.6	1.0	1.2	1.1
3.3 Interest on lending	0.0	0.0	0.0	0.1	0.0	0.0
3.4 Pension (old age + others)	9.4	1.8	14.3	3.8	3.7	7.4
3.5 Remittances from outside	8.9	2.4	6.8	0.7	4.3	4.7
3.6 Other receipts	0.3	0.8	0.1	5.8	0.3	1.4
4.0 Non-farm employment earnings	29.3	27.2	24.6	25.2	29.8	24.9
4.1 Govt./semi-govt. job	6.5	14.3	15.9	10.1	18.6	11.4
4.2 Large private company	11.0	6.4	6.1	14.3	9.7	8.7
4.3 Small business/industry	8.1	4.7	1.3	0.0	0.6	3.1
4.4 Family business/industry	1.0	0.0	0.9	0.8	0.0	0.6
4.5 Others	2.8	1.9	0.4	0.0	0.8	1.1
5. Self-employment earnings	11.3	21.5	14.5	11.9	15.7	13.5
6. Earnings from casual labour	12.3	2.3	0.3	0.0	34.5	6.0
7. Gross total per household	100.0	100.0	100.0	100.0	100.0	100.0
8. Gross total per capita per annum	7166	8033	15393	21875	5569	10923

Note: Figures are respective percentages of gross household income.
Source: Field Survey.

Within the classes of farm size, non-farm income was more important for marginal and small farmers who earned 55 to 71 per cent from agriculture and related activities, the remaining being contributed by non-farming activities. Large farmers, on the other hand, earned more than 90 per cent from farm and related activities. Livestock rearing was particularly significant for the marginal and small farmers as it contributed around 22 per cent of the earnings for the marginal farmers and 18 per cent for the small farmers, whereas large farmers had only 7 per cent of their total income from livestock. The importance of livestock was even more in the case of landless labourers who got around 30 per cent of their earnings from this particular source. Leasing out land also contributed around 5 per cent earnings for the landless labourers and 2 per cent for marginal farmers. Casual labour was a major source of earnings for the landless labourers although it also supplemented income for the marginal farmers. Self-employment contributed more than 10 per cent of the earnings in the case of marginal farmers and landless labourers, while for other classes its share was less than 2 per cent. Besides these activities, government and semi-government jobs also contributed around 8 to 9 per cent in the case of marginal, small and medium farmers.

Summing up the discussion on earnings in Punjab, farm and allied activities contributed 68-82 per cent of income in the case of marginal and small farmers and around 85-95 per cent in the case of medium and large farmers. Among the non-farming activities, self-employment, casual labour and government/private sector jobs were the major source of employment and earnings for the rural households in Punjab.

Bihar presents a completely different picture in terms of farm and non-farm earnings. The share of farm income in total household income was only 36 per cent in comparison to 76 per cent in the case of Punjab. The share of livestock activities was also very small, only 4 per cent whereas in Punjab it was above 8 per cent. Thus farming (and allied activities) in Bihar, although being full time occupations, the cultivators were not able to eke out even half of their earnings from them, forcing them to largely depend on non-farming activities to make both ends meet. Jobs in the government or private sector, self-employment, old age pensions, casual earnings and outside remittances were the major off-farm employment activities in Bihar. Across farm size categories, per household farm income was not as diverse as it was in Punjab. Share of farm business in total household income was 22 and 38 per cent respectively, for marginal and small farmers, while it was 47 per cent for large farmers. The share of livestock activities was almost same across all size classes, which averaged around 4 per cent. Regular jobs in the private or government sectors constituted around 15 to 20 per cent share in total household earnings across all size classes. The share of self-employment was also evenly distributed among different size classes while casual labour was the principal means of earning for landless labourers and marginal farmers alone.

Comparing gross household income in Punjab and Bihar, the statistics reveal that per household income in Punjab stood at more than twice that of Bihar. However, the difference was more than 4 times, in the case of farm business income. The off-farm income was higher in the case of Bihar in comparison to Punjab. Moreover, the difference in total household income between the two states was much lower in the lower strata of operated area as compared to higher strata among the selected households in the two states. It is more interesting to compare per capita income of the selected households in the two states. It is seen from the data that per capita income in Punjab was almost twice that of Bihar in the lower strata and more than twice in the upper strata of operational holdings. The average per capita per month earnings was Rs 2636 in Punjab and Rs 910 in Bihar.

Consumption Pattern

Table 5.10(a) and 5.10(b) provide statistics on consumption pattern of selected households in the two states. Figures in parentheses show that out of their total earnings, Punjab cultivators spent around 28 per cent on daily and durable consumer goods and services. In comparison, farming community in Bihar spent around 38 per cent of their total household income on consumer goods and services. The highest expenditure was incurred on food grains and non-food grains in both the states. Food grains include the expenditure incurred on cereals and pulses alone while non-food grains include expenditure on vegetables, milk, meat, fish, eggs, fruits (dry and fresh) etc.

In Punjab, around 11 per cent income was spent on food and non-food items. The percentage was highest in the case of landless labourers and marginal farmers and lowest in the case of large farmers. Marginal farmers spent more than 30 per cent of their income on food and non-food items while large farmers spent only 9 per cent on such items. Clothing, medical expenses and fuel and electricity were the other major expenses in the case of marginal farmers and landless labourers accounting for around 15 to 20 per cent income being spent on these items. Large farmers spent around 6 per cent on the above-mentioned components of consumption. After food, loan repayment on consumer loans was the biggest priority expenditure for the medium and large farmers. Despite increasing awareness, expenditure incurred on education was only 2.4 per cent of total income, and ranged between 2.3 in the case of marginal farmers to 4.3 among medium farmers. Overall, 60 per cent of the income of marginal farmers and 68 per cent of the income of landless labourers was spent on consumption goods, while large farmers spent only 27 per cent. Thus in the lower strata, farmers were left with very small margins to invest in productive uses in farming and other activities, whereas their counterpart bigger size farmers saved more than half of their income which could be used for investment purposes.

Table 5.10(a)

Consumption Expenditure: Punjab

(Rs per household per annum)

	Marginal	*Small*	*Medium*	*Large*	*Landless*	*Total*
Food grains	7667 (22.3)	8585 (20.2)	10344 (17.3)	17353 (13.5)	7204 (27.9)	10006 (18.0)
Non-food grains	9669 (28.2)	9625 (22.6)	11575 (19.3)	23570 (18.4)	6402 (24.8)	11743 (21.1)
Intoxicants	1007 (2.9)	1499 (3.5)	3659 (6.1)	5720 (4.5)	990 (3.8)	2398 (4.3)
Education	1316 (3.8)	3084 (7.2)	6331 (10.6)	13737 (10.7)	1175 (4.6)	4777 (8.6)
Fuel and electricity	3025 (8.8)	4853 (11.4)	6836 (11.4)	14159 (11.1)	2489 (9.6)	5961 (10.7)
Medical expenses	3177 (9.3)	2559 (6.0)	3542 (5.9)	6162 (4.8)	2614 (10.1)	3540 (6.4)
Transport	2216 (6.5)	2545 (6.0)	3361 (5.6)	13109 (10.2)	1131 (4.4)	4293 (7.7)
Clothing	3839 (11.2)	3966 (9.3)	6163 (10.3)	9297 (7.3)	2225 (8.6)	4799 (8.6)
Recreation including marriage etc.	916 (2.7)	998 (2.3)	1376 (2.3)	5796 (4.5)	559 (2.2)	1864 (3.4)
Loan repayment	1493 (4.3)	4840 (11.4)	6749 (11.3)	19177 (15.0)	1024 (4.0)	6245 (11.2)
Gross total	34325 (100.0)	42554 (100.0)	59936 (100.0)	128080 (100.0)	25813 (100)	55625 (100.0)

Note: Figures in parentheses are percentages of gross household consumption.

Source: Field Survey.

Table 5.10(b)

Consumption Expenditure: Bihar

(Rs per household per annum)

	Marginal	*Small*	*Medium*	*Large*	*Landless*	*Total*
Food grains	8204 (32.7)	9905 (26.6)	12283 (25.4)	14646 (19.4)	7050 (32.0)	9371 (27.7)
Non-food grains	6455 (25.7)	8986 (24.1)	12728 (26.3)	16272 (21.5)	4822 (21.9)	8251 (24.4)
Intoxicants	870 (3.5)	1025 (2.8)	1664 (3.4)	1924 (2.5)	774 (3.5)	1070 (3.2)
Education	798 (3.2)	3275 (8.8)	3203 (6.6)	12052 (16.0)	1352 (6.1)	2509 (7.4)
Fuel and electricity	900 (3.6)	1155 (3.1)	1693 (3.5)	2374 (3.1)	880 (4.0)	1162 (3.4)
Medical expenses	2148 (8.6)	3089 (8.3)	3368 (7.0)	5199 (6.9)	1419 (6.4)	2594 (7.7)
Transport	848 (3.4)	3293 (8.8)	2151 (4.4)	2525 (3.3)	1312 (6.0)	1731 (5.1)
Clothing	1969 (7.8)	2828 (2.6)	4344 (9.0)	11306 (15.0)	2144 (9.7)	3148 (9.3)
Recreation including marriage etc.	921 (3.7)	1587 (4.3)	2226 (4.6)	2932 (3.9)	1053 (4.8)	1405 (4.2)
Loan repayment	1411 (5.6)	2103 (5.6)	4532 (9.4)	6310 (8.4)	1217 (5.5)	2299 (6.8)
Others	573 (2.6)	0 (0.0)	150 (0.3)	0 (0.0)	0 (0.0)	255 (0.8)
Gross total	25099 (100.0)	37248 (100.0)	48341 (100.0)	75541 (100.0)	22021 (100.0)	33794 (100.0)

Note: Figures in parentheses are percentages of gross household consumption.
Source: Field Survey.

The consumption pattern in Bihar was almost similar to the pattern in Punjab. Food grains and non-food grains were the biggest items of consumption in Bihar as well. The proportion of income spent on food and non-food items among marginal and small farmers was above 25 per cent while large farmers spent only 12 per cent on these items. As in Punjab, in Bihar also clothing and medical expenses were accorded very high priority in consumption items. Loan repayment, education and transportation were the other chief items of consumption in Bihar. Finally, small and marginal farmers spent around 50 per cent of their income on consumption while large and medium farmers spent around 30 per cent on such items.

To sum up the findings, the figures on employment reveal the fact that whereas the maximum number of family members of large and medium farmers were absorbed in agriculture and allied activities, a comparatively higher number of family members in the case of marginal and landless labourers tried to find employment in allied and non-farm activities. In the crop sector, total labour absorption in harvesting, sowing and other activities in Punjab were calculated as 40 man-days per acre. In Bihar, on an average 43 man-days were employed in harvesting and sowing activities. Livestock rearing was a principal activity among the cultivators, and provided a major source of employment to the family labour in both the states. Comparing manpower in crop and livestock activities in these two states, around 59 per cent manpower was engaged in the crop sector and the remaining 41 per cent found employment in the livestock sector in Punjab. In Bihar, the ratio of crop and livestock employment was 54 and 46 per cent, respectively.

From the discussion on earnings from livestock, one can conclude that smaller size farmers supplemented their farming income through livestock activities much more than larger size farmers in both the states. Comparing gross household income in Punjab and Bihar, the statistics revealed that per household income in Punjab stood more than twice that of Bihar. Nonetheless, the difference was much higher, more than 4 times in the case of farm business income. Apparently, the off-farm income was much more for Bihar cultivators in comparison to Punjab. Moreover, the difference in total household income between the two states was much lower in the lower strata of operated area as compared to higher strata among the selected households in the two states. The results revealed that per capita income in Punjab was less than twice that of Bihar in the lower strata and almost twice in the upper strata of operational holdings. The consumption pattern in Bihar was almost similar to Punjab. Food grains and non-food grains were the biggest items of consumption in both the states. Clothing, fuel and electricity and medical expenses were accorded high priority among the consumption items. Loan repayment, education and transportation were the other chief items of consumption in both the states. Finally, small and marginal farmers spent around 50 to 60 per cent of their income on consumption, while large and medium farmers spent around 25 to 30 per cent on such items.

Annexure Table A-5.1

Total Man-Days Employed in Crop and Livestock Activities

(Man-days per household)

	Marginal	Small	Medium	Large	Landless	Total
Crop Sector						
Punjab						
Family labour	53.10	82.99	107.34	125.32	0.00	65.23
Attached labour	1.10	5.07	21.42	194.03	0.00	42.80
Hired labour	24.97	83.82	144.48	606.28	0.00	161.33
Total labour	79.18	171.88	273.24	925.63	0.00	269.35
Bihar						
Family labour	53.37	88.48	116.58	137.87	0.00	65.94
Attached labour	0.14	1.03	3.13	17.49	0.00	1.92
Hired labour	23.44	81.52	184.65	355.51	0.00	76.83
Total labour	76.96	171.03	304.36	510.87	0.00	144.68
Livestock Sector						
Punjab						
Family labour	143.5	169.1	166.8	161.9	121.1	148.8
Attached labour	5.0	21.4	50.8	109.8	3.2	34.9
Hired labour	1.7	1.6	6.2	15.0	0.0	4.4
Total labour	150.4	192.0	223.9	286.7	124.3	187.9
Bihar						
Family labour	175.5	101.7	105.5	83.4	27.5	118.0
Attached labour	1.3	1.2	0.0	28.8	0.0	2.7
Hired labour	0.1	0.9	1.7	8.3	1.5	1.3
Total labour	176.8	103.8	107.2	120.6	29.0	121.9

Source: Field Survey.

Annexure Table A-5.2

Man-Days Employed in the Crop and Livestock Activities

(Per cent)

	Marginal	*Small*	*Medium*	*Large*	*Landless*	*Total*
			Crop Sector			
Punjab						
Family labour	27.0	32.9	39.2	43.6	0.0	30.5
Attached labour	18.0	19.2	29.7	63.9	0.0	55.1
Hired labour	93.7	98.2	95.9	97.6	0.0	97.3
Total labour	34.5	47.2	55.0	76.4	0.0	58.9
Bihar						
Family labour	23.3	46.5	52.5	62.3	0.0	35.9
Attached labour	10.1	46.6	100.0	37.8	0.0	41.6
Hired labour	99.7	98.9	99.1	97.7	0.0	98.4
Total labour	30.3	62.2	74.0	80.9	0.0	54.3
			Livestock Sector			
Punjab						
Family labour	73.0	67.1	60.8	56.4	100.0	69.5
Attached labour	82.0	80.8	70.3	36.1	100.0	44.9
Hired labour	6.3	1.8	4.1	2.4	100.0	2.7
Total labour	65.5	52.8	45.0	23.6	100.0	41.1
Bihar						
Family labour	76.7	53.5	47.5	37.7	100.0	64.1
Attached labour	89.9	53.4	0.0	62.2	100.0	58.4
Hired labour	0.3	1.1	0.9	2.3	100.0	1.6
Total labour	69.7	37.8	26.0	19.1	100.0	45.7

Source: Field Survey.

Annexure Table A-5.3

Demographic Changes of Selected Households since 1991

(Per cent)

Category	% of Persons changed Occupation	Migrated to		Reasons of Migration					Nature of Migration		
		Rural	Urban	More Earning	Job/ Transfer	Tension in Village	Social Norms	Others	Temporary	Permanent	Circular
Punjab											
Marginal	2.0	100.0	0.0	87.5	0.0	0.0	0.0	12.5	25.0	37.5	37.5
Small	4.7	68.0	32.0	84.0	0.0	0.0	4.0	12.0	24.0	44.0	32.0
Medium	1.1	75.0	25.0	75.0	0.0	0.0	0.0	25.0	75.0	25.0	0.0
Large	1.4	60.0	40.0	100.0	0.0	0.0	0.0	0.0	50.0	30.0	20.0
Landless	0.5	100.0	0.0	100.0	0.0	0.0	0.0	0.0	33.3	0.0	66.7
Total	1.9	74.0	26.0	88.0	0.0	0.0	2.0	10.0	34.0	36.0	30.0
Bihar											
Marginal	10.1	33.0	67.0	74.3	10.9	1.0	2.0	11.9	70.3	22.8	6.9
Small	8.5	33.3	66.7	72.9	25.0	0.0	0.0	2.1	66.7	27.1	6.3
Medium	8.8	23.6	76.4	50.9	43.6	0.0	0.0	5.5	45.5	47.3	7.3
Large	4.9	22.2	77.8	44.4	33.3	0.0	0.0	22.2	55.6	11.1	33.3
Landless	8.9	44.1	55.9	75.0	15.6	0.0	0.0	9.4	59.4	31.3	9.4
Total	8.8	32.1	67.9	67.8	22.4	0.4	0.8	8.6	62.0	29.8	8.2

Source: Field Survey.

6 Globalisation in the Context of Punjab and Bihar

India now has about a decade's experience of living under the 'global umbrella'. The present study intends to look at the impact of the recent agriculture related policy changes on the emerging production, marketing, employment and earning status of small *versus* large farmers. The problems and stresses of switching over to an open economy are getting known. The idea is to see how marginal and small farmers, who together constitute the preponderant majority of cultivating households in rural India, and are likely to remain the backbone of Indian agriculture, are faring in all aspects of their economic existence in the post-reform years. There are reasons to believe that marginal and small farmers are likely to face unfavourable circumstances in their farm production and marketing on the one hand, and an unresponsive labour market on the other. It is essential, therefore, to see on the one hand, how marginal/small farmers in different parts of India are coping up with new production and marketing regimes, and how the various avenues of their employment and earnings, especially non-farm employment, are getting affected under the open market regimes.

In order to get an insight on the above issues, we incorporated a variety of qualitative and subjective questions in our questionnaire. In addition, objective information was sought on production, marketing, employment and various social factors for the current period as well as for the period before the reform process started. The year 1995 was taken as the starting point for the survey, as that was the period when the reform process just started in the agriculture sector. The information was based on the respondents' memory, as they generally did not keep any written records.

Here we undertake to summarise the comparison of the rural economies of Punjab and Bihar during the pre- and post-reform periods.

The chapter begins with a comparison of social indicators for the small and large farmers during these two periods. Section II presents land transactions during pre- and post-reforms, succeeded by Section III presenting the asset holdings by the households. Section IV makes a comparison of income and input use by small and large farmers during the two periods. Section V presents the major problems and difficulties faced by the selected households, and the last section summarises the findings.

Social Indicators

Housing, Electrification and Drinking Water

The incidence of sale and purchase of residential property in rural India is generally rare as in most of the cases such property moves automatically to the descendants through inheritance. The data also confirms this as 94 per cent of the rural houses in Punjab and 87 per cent in Bihar were inherited. The trend remained the same across various size classes and across the time period of last 15 years (Table 6.1). In the case of house ownership, the landless labourers in Bihar were at the lowest point with 30 per cent of them staying in rented-provided houses while all the remaining classes of households in both the states mostly owned the houses they were staying in. With respect to electrification of villages/ houses, around 93 per cent houses in Punjab were electrified, compared to around 41 per cent in Bihar. Electrification was added to around 15 per cent of the houses in Punjab and 10 per cent in Bihar during a period of 15 years after the reform process started in 1991. Electrification improved significantly for landless labourers from 50 per cent to 80 per cent and marginal farmers from 76 per cent to 92 per cent in Punjab. In Bihar, electrification increased from 20 per cent to 24 per cent in the case of landless labourers and from 23 to 35 per cent in the case of marginal farmers during this period. A majority of rural household depended on hand pumps for drinking water in both the states. Out of the selected households, 94 per cent in Punjab and 92 per cent in Bihar had hand pump as their main source of drinking water. In 1991, around 14 per cent of the households in Punjab and 16 per cent in Bihar depended on dug wells or ponds for drinking water. This ratio declined to 5 per cent in Punjab and 8 per cent in Bihar at the time of survey. Among various farm size categories, there were no major differences across various sizes in the case of source of drinking water.

Table 6.1

Changes in Social Indicators

(Percentage of households)

Category	*Year*	*Status of the House*				*Source of Drinking Water*		
		Inherited	*Bought*	*Others*	*Per cent Electrified*	*Hand pump*	*Dug Well*	*Pond/ River etc.*
Punjab								
Marginal	In 1991	95.3	3.5	1.2	75.6	84.9	8.1	7.0
	In 1995	96.5	2.3	1.2	80.2	88.4	4.7	7.0
	Current status	96.5	2.3	1.2	91.9	91.9	2.3	5.8
Small	In 1991	94.3	5.7	0.0	85.1	83.9	11.5	4.6
	In 1995	92.0	6.9	1.1	90.8	89.7	5.7	4.6
	Current status	92.0	6.9	1.1	95.4	96.6	0.0	3.4
Medium	In 1991	92.3	7.7	0.0	92.3	89.2	7.7	3.1
	In 1995	92.3	7.7	0.0	96.9	92.3	4.6	3.1
	Current status	92.3	7.7	0.0	100.0	98.5	0.0	1.5
Large	In 1991	100.0	0.0	0.0	100.0	96.6	2.3	1.1
	In 1995	100.0	0.0	0.0	100.0	97.7	1.1	1.1
	Current status	100.0	0.0	0.0	100.0	93.1	1.1	5.7
Landless	In 1991	90.1	5.9	4.0	49.5	76.2	5.9	17.8
	In 1995	90.1	5.9	4.0	67.3	82.2	5.0	12.9
	Current status	89.1	6.9	4.0	80.2	92.1	1.0	6.9
Total	In 1991	94.4	4.5	1.2	78.9	85.7	7.0	7.3
	In 1995	94.1	4.5	1.4	85.9	89.7	4.2	6.1
	Current status	93.9	4.7	1.4	92.7	94.1	0.9	4.9
Bihar								
Marginal	In 1991	90.3	3.4	6.3	22.9	82.9	8.0	9.1
	In 1995	90.3	3.4	6.3	30.3	89.1	3.4	7.4
	Current status	89.1	4.0	6.9	34.9	93.1	2.3	4.6
Small	In 1991	90.9	3.4	5.7	30.7	85.2	9.1	5.7
	In 1995	90.9	2.3	6.8	42.0	90.9	5.7	3.4
	Current status	89.8	3.4	6.8	43.2	94.3	3.4	2.3
Medium	In 1991	96.8	3.2	0.0	45.2	87.1	9.7	3.2
	In 1995	96.8	3.2	0.0	54.8	88.7	8.1	3.2
	Current status	96.8	3.2	0.0	56.5	90.3	6.5	3.2
Large	In 1991	100.0	0.0	0.0	72.4	93.1	0.0	6.9
	In 1995	100.0	0.0	0.0	75.9	96.6	0.0	3.4
	Current status	100.0	0.0	0.0	79.3	96.6	0.0	3.4
Landless	In 1991	65.8	2.6	31.6	19.7	77.6	14.5	7.9
	In 1995	65.8	2.6	31.6	22.4	81.6	10.5	7.9
	Current status	67.1	2.6	30.3	23.7	86.8	7.9	5.3
Total	In 1991	87.7	3.0	9.3	30.5	83.7	9.1	7.2
	In 1995	87.7	2.8	9.5	37.9	88.6	5.6	5.8
	Current status	87.2	3.3	9.5	40.7	92.1	4.0	4.0

Source: Field Survey.

Place of Defecation

A major shift has taken place from open field defecation to pit latrine and septic tank latrines in Punjab during the last 15 years. In Bihar, on the other hand, open fields still remain the most common place for defecation. In Punjab currently 25 per cent of the households were using pit latrine while this proportion was only 5 per cent in 1991. Similarly, septic tank users also increased from 5 per cent to 16 per cent during this period while people using open fields declined from 87 per cent to 53 per cent. A major shift from open fields to manufactured latrines took place in the case of large and medium farmers. Twenty-four per cent of the large farmers and 34 per cent of the medium farmers were using open fields for defecation at the time of the survey. In the case of landless labourers, marginal farmers and small farmers the ratio was 84, 61 and 51 per cent, respectively. In Bihar on the other hand no major shift from open field defecation was observed during the last 15 years. The percentage of households using septic tanks or pit latrines increased from 23 per cent in 1991 to 27 per cent during the current period. However, the percentage of households using septic tanks was very high, around 59 per cent in the case of large farmers and 45 per cent in the case of medium farmers. The percentage was very low in the case of marginal farmers and landless labourers, which averaged around 14 per cent. Thus, one concludes that during the last 15 years a major chunk of households in all size classes shifted from open field defecation to manufactured latrines in Punjab while in Bihar the open field trend continues to dominate, revealing the sorry state of health standards in that state.

Fuel and Firewood

It is seen from Table 6.3 that wood or (coal) and cow dung still remain the major sources of fuel among the selected households in both the states. More scientific means like *gobar* gas plants and LPG were being used by a scanty number of households, less than 5 per cent in both the states. In Punjab, almost all categories were rearing animals like cows and buffaloes. Therefore, cow dung cake was the main fuel used by a majority of households. Bihar on the other hand being a centre of coal, the latter was a main fuel for the majority of the households. These two sources together were the main source of fuel for 93 per cent of the households in Punjab and 85 per cent in Bihar. Hay, leaves and agricultural waste were the other fuels used by around 4 per cent of the households in Punjab and 10 per cent in Bihar. However, LPG was gaining popularity among the large households in both these states. As LPG involved regular expenditure for refilling and *gobar* gas required huge initial investment, only very rich farmers were turning towards these sources for fuel.

Table 6.2

Place of Defecation of the Selected Households

(Percentage of households)

		Septic Tank	*Pit Latrine*	*Covered Dry*	*Open Field*
Punjab					
Marginal	In 1991	3.5	1.2	1.2	94.2
	In 1995	3.5	5.8	3.5	87.2
	Current status	11.6	20.9	7.0	60.5
Small	In 1991	2.3	2.3	2.3	93.1
	In 1995	4.6	10.3	5.7	79.3
	Current status	17.2	25.3	6.9	50.6
Medium	In 1991	3.1	10.8	6.2	80.0
	In 1995	7.7	24.6	12.3	55.4
	Current status	18.5	35.4	12.3	33.8
Large	In 1991	12.6	14.9	5.7	66.7
	In 1995	18.4	20.7	9.2	51.7
	Current status	31.0	34.5	10.3	24.1
Landless	In 1991	3.0	0.0	0.0	97.0
	In 1995	3.0	2.0	1.0	94.1
	Current status	3.0	11.9	1.0	84.2
Total	In 1991	4.9	5.4	2.8	86.9
	In 1995	7.3	11.7	5.9	75.1
	Current status	15.7	24.6	7.0	52.6
Bihar					
Marginal	In 1991	10.3	2.3	1.1	86.2
	In 1995	10.3	2.3	1.7	85.6
	Current status	13.8	2.3	1.7	82.2
Small	In 1991	26.4	1.1	1.1	71.3
	In 1995	29.9	1.1	0.0	69.0
	Current status	32.2	1.1	0.0	66.7
Medium	In 1991	38.7	3.2	6.5	51.6
	In 1995	43.5	3.2	3.2	50.0
	Current status	45.2	1.6	3.2	50.0
Large	In 1991	48.3	3.4	17.2	31.0
	In 1995	51.7	3.4	17.2	27.6
	Current status	58.6	3.4	13.8	24.1
Landless	In 1991	14.9	0.0	1.4	83.8
	In 1995	14.9	0.0	1.4	83.8
	Current status	14.9	0.0	1.4	83.8
Total	In 1991	21.1	1.9	3.1	73.9
	In 1995	22.8	1.9	2.6	72.8
	Current status	25.4	1.6	2.3	70.7

Source: Field Survey.

Table 6.3

Main Source of Fuel of the Selected Households

(Percentage of households)

		Wood/ Coal	Hay/ Leaves	Cow Dung Cake	Agricultural Waste	Gobar Gas Plant	LPG
Punjab							
Marginal	In 1991	44.2	0.0	54.7	1.2	0.0	0.0
	In 1995	44.2	0.0	54.7	1.2	0.0	0.0
	Current status	43.0	0.0	54.7	1.2	0.0	1.2
Small	In 1991	41.4	4.6	52.9	0.0	0.0	1.1
	In 1995	42.5	4.6	51.7	0.0	0.0	1.1
	Current status	39.1	4.6	51.7	0.0	0.0	4.6
Medium	In 1991	43.1	0.0	55.4	1.5	0.0	0.0
	In 1995	38.5	0.0	55.4	3.1	1.5	1.5
	Current status	38.5	0.0	52.3	6.2	1.5	1.5
Large	In 1991	42.5	0.0	48.3	4.6	0.0	4.6
	In 1995	40.2	1.1	46.0	5.7	0.0	6.9
	Current status	37.9	0.0	46.0	5.7	1.1	9.2
Landless	In 1991	52.5	4.0	43.6	0.0	0.0	0.0
	In 1995	52.5	4.0	43.6	0.0	0.0	0.0
	Current status	52.5	3.0	44.6	0.0	0.0	0.0
Total	In 1991	45.1	1.9	50.5	1.4	0.0	1.2
	In 1995	44.1	2.1	49.8	1.9	0.2	1.9
	Current status	42.7	1.6	49.5	2.3	0.5	3.3
Bihar							
Marginal	In 1991	58.6	3.4	27.6	9.8	0.0	0.6
	In 1995	59.8	2.3	27.6	9.8	0.0	0.6
	Current status	58.6	2.3	28.2	9.8	0.0	1.1
Small	In 1991	54.7	0.0	38.4	4.7	0.0	2.3
	In 1995	57.0	0.0	36.0	4.7	0.0	2.3
	Current status	57.0	0.0	34.9	4.7	0.0	3.5
Medium	In 1991	51.6	0.0	38.7	3.2	0.0	6.5
	In 1995	51.6	0.0	37.1	3.2	0.0	8.1
	Current status	50.0	0.0	37.1	3.2	0.0	9.7
Large	In 1991	48.3	0.0	31.0	3.4	3.4	13.8
	In 1995	48.3	0.0	27.6	6.9	3.4	13.8
	Current status	48.3	0.0	24.1	6.9	3.4	17.2
Landless	In 1991	66.2	17.6	13.5	2.7	0.0	0.0
	In 1995	67.6	17.6	12.2	2.7	0.0	0.0
	Current status	64.9	16.2	12.2	2.7	0.0	4.1
Total	In 1991	57.4	4.5	29.2	6.1	0.2	2.6
	In 1995	58.6	4.0	28.0	6.4	0.2	2.8
	Current status	57.4	3.8	27.8	6.4	0.2	4.5

Source: Field Survey.

Land Transactions

It has been pointed out in recent literature (Johl and Ray, 2002) that a number of small and marginal farmers in Punjab are leasing out or selling off their land and leaving the agriculture business because of rising cost and stagnating productivity in farming. In our previous chapter we saw that our data vindicated the hypothesis of switching of tenancy by the marginal farmers in favour of large farmers. Statistics on land transactions would reveal whether switching is a short time phenomenon or whether it is a permanent feature.

It is seen from Table 6.4 that all household categories in both the states had sold or bought some land during the pre-reform and reform phase periods. However, the difference in sale and purchase of land during these two periods within each category was only marginal. In Punjab, on an average, 0.10 acres of land per household were sold before 1995, while after the reform process, 0.14 acres per household were sold by the selected households. On the other hand, the land bought in the pre- and reform periods was 0.18 and 0.12 acres per household respectively. Thus, comparatively higher land was sold and less was purchased during the reform period than in the pre-reform period in Punjab. However, the difference in transacted land during these two periods was only nominal. In Bihar, on the other hand, land sold and bought during the reform period was a little higher than that of the pre-reform period. In the pre-reform period 0.04 and 0.05 acres per household, respectively were bought and sold while during the reform period both the figures were a little higher at 0.06 and 0.08 acres per household, respectively.

Comparing the combined case of small, marginal and landless cultivators (small) *versus* large and medium farmers (large), it is seen from the statistics that in Bihar, both large and small farmers sold as well as bought more land during the reform period compared to the pre-reform period. The results were not as unambiguous in Punjab where small farmers sold 0.27 acres per household in the pre-reform period and 0.47 acres per household during the reform period. Large farmers on the other hand sold 0.30 acres per household in the pre-reform period and 0.27 during the reform period. This trend verifies the earlier assertion that in post liberalisation era, the smaller categories are moving away from farming in Punjab. However, the statistics on land bought does not reveal the same trend. The smaller size farmers bought 0.10 acres per household in the pre-reform period while they bought 0.29 acres per household during the reform period. Larger size realised an opposite trend as they bought higher amount i.e., 0.85 acres per household in the pre-reform period and only 0.30 acres per household during the reform period. The above results do not support the inference that smaller size holders in Punjab are selling out land in favour of large size farmers. Thus, small size farmers might have been leasing out land but they did not give up their right on the land by selling it off. In the given circumstances, it makes a perfect case for liberalisation of tenancy laws in the state. Allowing legal tenancy in the state would be in the interest of farming community in general.

Table 6.4

Details of Land Transactions

	Land Sold (Acres per HH)		*Land Bought (Acres per HH)*		*Land Sold (Rs per Acre)*		*Land Bought (Rs per Acre)*	
	Before 1995	*After 1995*	*Before 1995*	*After 1995*	*Before 1995*	*After 1995*	*Before 1995*	*After 1995*
Punjab								
Marginal	0.19	0.09	0.01	0.02	115000	93800	175000	152500
Small	0.06	0.31	0.07	0.23	71667	190227	77000	186125
Medium	0.14	0.23	0.16	0.03	155000	196000	59288	180000
Large	0.16	0.04	0.69	0.27	174500	325000	168500	216786
Landless	0.02	0.07	0.02	0.04	126667	338571	47333	303333
Total	0.10	0.14	0.18	0.12	132364	219500	129025	210310
Bihar								
Marginal	0.03	0.02	0.04	0.11	162727	144091	84750	130987
Small	0.02	0.11	0.10	0.07	108357	131462	161448	159817
Medium	0.03	0.10	0.07	0.06	100000	167056	129167	184857
Large	0.17	0.12	0.13	0.22	118750	162500	66600	154200
Landless	0.02	0.03	0.00	0.01	146250	145000	0	145000
Total	0.04	0.06	0.05	0.08	129106	147012	112159	152361

Source: Field Survey.

Farm Investment in Asset Holdings

The market value of farm assets gives us an idea about the cumulative investments made by the farmers in the farm business. To an extent these farm assets are an important determinant of land productivity. Tables 6.5(a) and 6.5(b) present group-wise data on per household and per acre assets other than landholdings which include farm machinery and equipment, i.e., tractors, trolleys, harvesters, threshers, pump sets, engines, etc.; buildings used for keeping livestock and storage bins for keeping grains, during the pre-reform and reform period. As expected, there was a positive relationship between per household assets and farm size in both the states. However, the positive association did not hold in the case of per acre asset holdings. Small and marginal farmer invested a much higher amount per acre compared to large farmers in both the states. The value of assets per acre increased for medium size farmers and declined for the large farmers in Punjab. In Bihar, marginal farmers' per acre investment was much higher compared to all other farm sizes.

On an average, Punjab farmers invested a much higher amount on farming compared to Bihar, which was also reflected in their land productivity. Per household investment in Punjab averaged around Rs 65,000 whereas the average value in Bihar was only Rs 16,000. In terms of per acre investment, Punjab farmers invested around double the amount of that of the Bihar farmers. Comparing pre-reform with the reform period, our data support the much discussed (Chand and Kumar, 2004) all-India phenomenon of rapidly increasing private investment in agriculture even with declining public investment in the nineties. It is seen from the tables that farm investment in Punjab increased by 78 per cent from the pre-reform to the reform period. The increase was even higher in the case of Bihar where investment increased by 125 per cent during the reform periods. Moreover, the bigger proportion of increase in investment came through small and medium farmers in Punjab and through marginal farmers in Bihar. Therefore, the above statistics nullify the apprehensions expressed by several quarters (Bhalla, 2005) that the small and marginal farmers would be left behind in the process of liberalisation and globalisation.

Accounting for assets in which farmers invested the highest amount, it is seen from the tables that tractors were the biggest component of investment in both the states. A huge amount per household was spent on submersible tube wells in Punjab although it did not create any new asset and was just to supplement the earlier investment made in mono-block

Table 6.5(a)

Ownership of Productive Assets Before and During Reform Phase

(Rs per household)

Category	Marginal		Small		Medium		Large		Landless		Total	
	Before 1995	*After 1995*	*Before 1995*	*After 1995*	*Before 1995*	*After 1995*	*Before 1995*	*After 1995*	*Before 1995*	*After 1995*	*Befo 1995*	*After 1995*
Punjab												
Tractor	7208	6364	10171	23244	26517	73086	46929	115282	2097	4637	16845	39927
Trolley	649	727	1159	2805	7103	7397	11776	9906	371	927	3765	3925
Tractor Implements	335	404	551	1204	2053	3447	3927	4604	95	149	1257	1736
Thresher	0	0	366	463	638	2621	741	2094	0	0	305	864
Combine	0	1623	0	0	0	0	82	3659	0	0	16	1024
Pump set (sub)	84	2610	313	4963	1948	11638	3571	26596	468	968	1189	8600
Pump set (mono)	1301	964	2020	4878	2252	2284	3976	7000	194	484	1780	2962
Pump (diesel)	377	156	817	468	1698	543	2106	682	512	85	1026	353
Sprayer	21	35	34	64	80	170	158	216	16	13	57	89
Storage bin	216	246	443	423	596	526	1164	846	112	90	470	393
Sheds (cattle)	4442	2701	4073	5061	7276	3534	14112	4400	1556	1177	5846	3164
Others	1288	806	2198	1195	2817	2514	13671	5367	433	301	3893	1876
Total	15920	16637	22145	44768	52978	107760	102215	180653	5853	8832	36452	64913
% change in total assets	(4.5)		(102.2)		(103.4)		(76.7)		(50.9)		(78.1)	

contd...

...contd...

Category	*Marginal*		*Small*		*Medium*		*Large*		*Landless*		*Total*	
	Before 1995	*After 1995*	*Before 1995*	*After 1995*	*Before 1995*	*After 1995*	*Before 1995*	*After 1995*	*Before 1995*	*After 1995*	*Befo 1995*	*After 1995*
Bihar												
Tractor	1714	7600	3693	7670	5887	16532	29828	60517	0	4605	4314	11942
Trolley	114	726	420	1477	645	1048	1552	7724	0	789	330	1409
Tractor Implements	39	67	94	125	64	467	351	1359	0	243	68	255
Thresher	80	400	283	1000	1319	2194	1403	6483	66	0	387	1121
Combine	0	0	0	0	0	0	0	152	0	0	0	10
Pump set (sub)	-	-	-	-	-	-	-	-	-	-	-	-
Pump set (mono)	0	17	102	0	315	0	69	207	0	0	71	21
Pump (diesel)	842	437	1958	1176	3382	1339	5578	2131	368	211	1672	793
Sprayer	0	19	35	0	73	0	62	103	26	33	27	21
Storage bin	0	10	7	11	6	0	0	0	3	0	3	6
Sheds (cattle)	0	57	18	0	0	0	0	0	0	789	4	163
Others	203	464	391	292	550	512	850	352	66	437	311	423
Total	2993	9797	7002	11752	12243	22092	39692	79028	530	7108	7187	16164
% change in total assets	(227.3)		(67.8)		(80.4)		(99.1)		(124.1)		(124.9)	

Source: Field Survey.

pump sets due to fall in water table, as was also highlighted by many researchers in the past (Vaidyanathan, 2006; 1996; 2001). Tractor implements, e.g., trolley, thresher, combine etc., were the other assets in which large investments were made by the farmers in both the states. It

Table 6.5(b)

Ownership of Productive Assets Before and During Reform Phase

(Rs per acre)

	Marginal		*Small*		*Medium*		*Large*		*Total*	
	Before 1995	*After 1995*	*Before 1995*	*After 1995*	*Before 1995*	*After 1995*	*Before 1995*	*After 1995*	*Before 1995*	*After 1995*
Punjab										
Tractor	4503	3975	2568	5868	3441	9485	2056	5050	2531	5999
Trolley	406	454	292	708	922	960	516	434	566	590
Tractor implements	209	252	139	304	266	447	172	202	189	261
Thresher	0	0	92	117	83	340	32	92	46	130
Combine	0	1014	0	0	0	0	4	160	2	154
Pump set (sub)	53	1631	79	1253	253	1510	156	1165	179	1292
Pump set (mono)	813	602	510	1231	292	296	174	307	267	445
Pump set (diesel)	235	97	206	118	220	70	92	30	154	53
Sprayer	13	22	9	16	10	22	7	9	9	13
Storage bin	135	154	112	107	77	68	51	37	71	59
Sheds (cattle etc.)	2775	1687	1028	1278	944	459	618	193	878	475
Others	804	503	555	302	366	326	599	235	585	282
Total	9945	10393	5590	11302	6875	13984	4478	7914	5477	9753
Bihar										
Tractor	1460	6474	1003	2083	809	2273	1858	3771	1283	3552
Trolley	97	618	114	401	89	144	97	481	98	419
Tractor implements	33	57	26	34	9	64	22	85	20	76
Thresher	68	341	77	272	181	302	87	404	115	333
Combine	0	0	0	0	0	0	0	9	0	3
Pump set (sub)	-	-	-	-	-	-	-	-	-	-
Pump set (mono)	0	15	28	0	43	0	4	13	21	6
Pump set (diesel)	717	372	532	319	465	184	348	133	497	236
Sprayer	0	17	10	0	10	0	4	6	8	6
Storage bin	0	9	2	3	1	0	0	0	1	2
Sheds (cattle etc.)	0	49	5	0	0	0	0	0	1	48
Others	173	395	106	79	76	70	53	22	93	126
Total	2549	8345	1902	3192	1683	3038	2473	4924	2137	4807

Source: Field Survey.

was strange to observe that investment in electric or diesel pump sets in Bihar came down during the reform period. It appears that due to non-electrification of villages and lack of quality electricity where wiring is available, farmers are not making new investments in the bore wells. Similarly, investment on sheds and storage also came down during the reform period. The trend of investment was almost similar for all categories in both the states except the case of marginal farmers in Punjab. In their case it is apparent that these farmers, in the wake of rising cost of operation, were gradually leaving farming and hardly making any new investment on the farm. The asset holdings declined for almost all items in their case with the exception of submersible pump sets and combines, which might have been over reporting by one or two individuals as investment in these items involved a huge amount of expenses.

Ownership of Livestock

Animal husbandry is an integral part of agriculture, and not only provides employment to the rural population during the off-season but also supplements their farming income. Almost every household in our sample kept some animals, predominant among them for the milch purposes and a few for draught and meat purposes. Although large scale tractorisation (especially in Punjab) has considerably reduced the role of animal power in agriculture, yet draught animals are kept for daily transport of fodder and light and sundry work. It is seen from the table that the per household value of draught animals in Punjab was as low as Rs 566, and on an average the number of draught animals kept was only 0.30 in contrast to the pair of bullocks kept earlier for the purpose of ploughing. In Bihar, per household value of draught animals was twice that in Punjab.

Cows and buffaloes were the main milch animals. However, buffalo milk is preferred in the Indian sub-continent against the trend of cow milk the world over. The same trend was also reflected by our sample data. The numbers as well as value of animals kept by the households was the maximum in the case of buffaloes in both the states. In Punjab, on an average two buffaloes were kept by each household while in Bihar the average number was 0.58. Because of higher productivity, cross-bred cows were more popular compared to local breed (*desi*) cows. Per household value of crossbred cows was Rs 2,978 compared to Rs 781 of *desi* cows in Punjab.

Table 6.6

Ownership of Livestock Before and During Reforms

		Cow (Desi)	*Cow (Hybrid)*	*Female Buffalo*	*Female Goat*	*Young Cattle*	*Draught Animal*	*Poultry*	*Other Animals*
Punjab									
Marginal	No. before	0.3	0.2	1.5	0.0	0.4	0.1	0.0	0.0
	No after	0.3	0.4	1.4	0.1	0.9	0.1	0.0	0.1
	Rs. per hh	1184	2649	13466	171	941	160	0	279
	Rs./animal	4343	7286	9970	2640	993	1367	0	5375
Small	No. before	0.3	0.3	2.1	0.0	0.9	0.3	0.0	0.0
	No after	0.3	0.3	2.3	0.0	1.1	0.4	0.0	0.0
	Rs. per hh	521	2689	23380	18	1078	752	18	73
	Rs./animal	2033	7875	10363	1500	983	1763	1500	6000
Medium	No. before	0.1	0.4	2.3	0.1	0.8	0.3	0.0	0.0
	No after	0.1	0.6	2.6	0.2	1.8	0.4	0.0	0.0
	Rs. per hh	121	6000	28133	241	1973	952	0	198
	Rs./animal	1400	9943	10735	1167	1100	2400	0	5750
Large	No. before	0.3	0.6	4.1	0.0	2.1	0.5	0.0	0.0
	No after	0.3	0.6	3.8	0.0	2.6	0.6	0.0	0.0
	Rs. per hh	1154	4884	42999	18	3845	1188	0	353
	Rs./animal	4459	7983	11315	1500	1472	1980	0	15000
Landless	No. before	0.2	0.0	1.0	0.6	0.3	0.0	0.0	0.0
	No after	0.2	0.1	1.0	0.6	0.6	0.1	0.0	0.0
	Rs. per hh	754	653	8212	836	693	90	0	60
	Rs./animal	3596	7364	8630	1329	1177	1388	0	1500
Total	No. before	0.2	0.3	2.1	0.2	0.9	0.2	0.0	0.0
	No after	0.2	0.4	2.1	0.2	1.3	0.3	0.0	0.0
	Rs. per hh	781	2978	21735	314	1615	566	4	306
	Rs./animal	3500	8238	10498	1380	1224	1915	1500	9321
Bihar									
Marginal	No. before	0.22	0.08	0.36	0.09	0.15	0.19	0.00	0.07
	No after	0.29	0.11	0.49	0.14	0.35	0.16	0.02	0.07
	Rs. per hh	1362	1034	4257	64	499	669	69	194
	Rs./animal	4675	9526	8663	465	1431	4182	3030	2615
Small	No. before	0.33	0.19	0.64	0.06	0.36	0.42	0.00	0.39
	No after	0.22	0.23	0.80	0.07	0.39	0.48	0.00	0.15
	Rs. per hh	1315	1892	6324	240	538	1983	0.00	520
	Rs./animal	6089	8325	7950	3517	1391	4155	0	3519
Medium	No. before	0.26	0.18	0.79	0.02	0.32	0.61	0.00	0.21
	No after	0.31	0.39	0.84	0.08	0.55	0.50	0.00	0.13
	Rs. per hh	1465	3815	7258	495	913	2911	0	968
	Rs./animal	4779	9854	8654	6140	1665	5823	0	7500
Large	No. before	0.52	0.83	1.14	0.00	0.66	0.83	0.00	0.17
	No after	0.45	0.59	1.03	0.00	0.52	0.28	0.00	0.07
	Rs. per hh	2455	5966	7103	0	800	1241	0	414
	Rs./animal	5477	10176	6867	0	1547	4500	0	6000
Landless	No. before	0.14	0.09	0.09	0.07	0.13	0.05	0.00	0.04
	No after	0.12	0.09	0.17	0.11	0.17	0.01	0.00	0.12
	Rs. per hh	599	711	1242	83	247	7	0	92
	Rs./animal	5056	7714	7262	788	1446	500	0	778
Total	No. before	0.25	0.17	0.48	0.06	0.25	0.32	0.00	0.16
	No after	0.26	0.20	0.58	0.10	0.37	0.26	0.01	0.10
	Rs. per hh	1306	1886	4772	161	542	1183	28	369
	Rs./animal	5060	9322	8175	1610	1485	4624	3030	3528

Source: Field Survey.

In Bihar, per household value of cross-bred and *desi* cows was Rs 1,886 and Rs 1,306, respectively.

Comparing farm size categories, although percentage share of dairy in total income was highest for marginal and small farmers in both the states, the number of animals kept was highest in the case of large farmers, in Punjab as well as in Bihar. The large farmers in Punjab had around 4 buffaloes per household compared to around 2 in the case of small and marginal farmers. In Bihar the average number was around one for large farmers and 0.50 for marginal farmers. However, the large farmers having higher size of family and hired labour had a much higher milk requirements and that might be the reason that smaller size farmers were earning more by selling milk, compared to larger size farmers who were keeping a higher proportion of their production for family consumption. Comparing the two phases, no significant differences were discernible in the ownership of number of livestock during these two periods.

Ownership of Household Assets

Ownership of superior assets is indicative of the social status of the household. Among the selected households, the major assets identified were house, television (coloured or black and white (B&W)), video-cassette recorder (VCR)/video cassette player (VCP), two/four wheelers, air-conditioner, LPG, microwave etc. With the beginning of liberalisation in India, there has been spurt in demand for electronic assets as a result of: (i) rising per capita income/earning opportunities; (ii) availability of these commodities at cheaper rate due to reduced excise and custom duties and (iii) increasing demonstration effect among the consumers. The spurt is clearly visible in the urban living standards and presumably is present in the rural areas as well. We tried to capture the change in living standards among our selected households in order to see whether it is present among only the affluent or is a general phenomenon.

Table 6.7 presents the statistics on changing status of ownership of assets by selected households during the reform period. A look at the data reveals that ownership of almost all these assets doubled in Punjab during the reform period. Barring housing, all other assets realised a 50 to 100 per cent increase during the pre-reform and the reform phase. The number of

Table 6.7

Ownership of Other Household Assets during the Two Phases

(No. and Rs per household)

Category		House	Television	VCR/ VCP	Two Wheeler	Four Wheeler	AC	LPG	Others
Punjab									
Marginal	No. before 1995	1.01	0.43	0.03	0.08	0.01	0.05	0.17	10.30
	No. after 1995	1.03	0.68	0.03	0.16	0.03	0.19	0.52	15.62
	% change	(1.3)	(57.6)	(0.0)	(100.0)	(100.0)	(275.0)	(207.7)	(51.7)
	Value	130779	2338	143	1621	2208	174	1137	7840
Small	No. before 1995	1.01	0.48	0.01	0.22	0.05	0.16	0.17	13.05
	No. after 1995	1.05	0.85	0.02	0.44	0.07	0.37	0.61	19.40
	% change	(3.6)	(79.5)	(100.0)	(100.0)	(50.0)	(130.8)	(257.1)	(48.7)
	Value	202427	3316	135	6083	5366	480	1292	9348
Medium	No. before 1995	1.07	0.64	0.07	0.38	0.02	0.17	0.34	16.45
	No. after 1995	1.14	0.93	0.12	0.64	0.09	0.47	0.81	23.21
	% change	(6.5)	(45.9)	(75.0)	(68.2)	(400.0)	(170.0)	(135.0)	(41.1)
	Value	263147	4521	584	8476	11552	736	2107	13105
Large	No. before 1995	1.06	0.74	0.16	0.65	0.20	0.59	0.45	22.55
	No. after 1995	1.14	1.13	0.21	1.01	0.33	1.13	0.89	31.67
	% change	(7.8)	(52.4)	(28.6)	(56.4)	(64.7)	(92.0)	(100.0)	(40.4)
	Value	540294	7406	1299	16355	54220	7655	2255	29957
Landless	No. before 1995	0.98	0.22	0.00	0.02	0.00	0.00	0.01	7.28
	No. after 1995	1.01	0.54	0.01	0.06	0.01	0.05	0.18	10.58
	% change	(2.5)	(148.1)	-	(166.7)	-	-	(2100.0)	(45.3)
	Value	86286	940	20	834	1202	58	305	4025
Total	No. before 1995	1.02	0.47	0.05	0.24	0.05	0.18	0.20	13.23
	No. after 1995	1.06	0.80	0.07	0.42	0.10	0.41	0.55	19.12
	% change	(4.1)	(70.4)	(42.9)	(72.1)	(82.6)	(126.0)	(173.3)	(44.5)
	Value	231352	3428	397	6124	14173	1769	1280	12150

contd...

...contd...

Category		*House*	*Television*	*VCR/ VCP*	*Two Wheeler*	*Four Wheeler*	*AC*	*LPG*	*Others*
Bihar									
Marginal	no before 1995	1.01	0.15	0.01	0.03	0.01	0.00	0.06	3.97
	No after 1995	1.07	0.21	0.01	0.03	0.01	0.00	0.07	7.07
	% change	(6.8)	(42.3)	(0.0)	(0.0)	(0.0)	-	(9.1)	(78.1)
	Value	78238	385	24	630	286	0	110	4389
Small	no before 1995	1.02	0.25	0.00	0.10	0.03	0.01	0.10	5.18
	No after 1995	1.26	0.25	0.00	0.14	0.03	0.01	0.10	10.24
	% change	(23.3)	(0.0)	-	(33.3)	(0.0)	(0.0)	(0.0)	(97.6)
	Value	105755	802	0	2193	5170	0	184	10268
Medium	no before 1995	0.98	0.48	0.02	0.27	0.02	0.02	0.29	6.15
	No after 1995	1.06	0.52	0.02	0.31	0.02	0.02	0.29	17.24
	% change	(8.2)	(6.7)	(0.0)	(11.8)	(0.0)	(0.0)	(0.0)	(180.6)
	Value	151008	1873	161	5202	2419	44	704	15024
Large	no before 1995	1.00	0.66	0.00	0.31	0.17	0.00	0.41	7.14
	No after 1995	1.38	0.62	0.00	0.38	0.17	0.00	0.45	20.90
	% change	(37.9)	-(5.3)	-	(22.2)	(0.0)	-	(8.3)	(192.8)
	Value	257241	2117	0	7683	14483	0	969	16972
Landless	no before 1995	0.91	0.12	0.01	0.07	0.01	0.00	0.11	2.99
	No after 1995	0.95	0.13	0.01	0.09	0.01	0.00	0.12	5.68
	% change	(4.3)	(11.1)	(0.0)	(40.0)	(0.0)	-	(12.5)	(90.3)
	Value	57471	605	38	721	4013	0	206	4100
Total	no before 1995	0.99	0.25	0.01	0.11	0.03	0.00	0.13	4.57
	No after 1995	1.11	0.28	0.01	0.13	0.03	0.00	0.14	9.87
	% change	(12.2)	(12.3)	(0.0)	(19.6)	(0.0)	(0.0)	(5.2)	(116.0)
	Value	102764	841	40	2101	3209	6	286	7923

Source: Field Survey.

households owning a television, two and four wheelers, air conditioners and LPG gas doubled during this period in Punjab. In Bihar, due to lower productivity, subsistence nature of agriculture and lack of earning opportunities elsewhere, no such spurt in consumer goods was observed. The data on percentage change in assets ownership shows only a nominal increase, less than 10 per cent, in number of household assets in Bihar during this period. There was hardly any one among the selected households in the state owning VCR/VCPs, four-wheelers or air conditioners. Comparing different farm size categories, it is observed from the table that in Punjab, small and marginal farmers had an equal rise in their living standard to their counterpart large farmers. In Bihar, on the other hand, only large farmers observed a major increase in ownership of houses, two wheelers and a few other household assets not matched by their counterpart small and marginal farmers.

Households Views on Effect of Globalisation on Rural Economy

The objective of quantitative analysis is to quantify the effect of ongoing changing government policies, and opening up of the overall global environment on the living standard of the respondents. A variety of questions related to economic and social aspects were posed to them and their replies are summarised in the following tables.

Major Problems Faced by the Respondents

In response to our survey on major problems faced, the respondents identified health, poverty, basic facilities, agriculture, employment and litigation as the major problems faced by them. Health problems and the lack of availability of suitable health centres or hospitals and doctors at affordable rates was the biggest problem faced by the households in both the states. Around 40 per cent of the households in Punjab and 55 per cent in Bihar cited health problems as their biggest malaise, and the small and marginal farmers were not the only to suffer but the large farmers as well.

Shortage of basic facilities, stagnating land productivity and poverty or lack of money, especially during the slack season due to lack of supplementary earnings were the other major problems among the selected farmers in the two states (Table 6.8). To our question on whether the households get sufficient food for the whole year, almost all selected households gave a positive reply, indicating no mass hunger among them not only in Punjab but also in Bihar, which is also one of the most poverty-ridden states in the country.

Table 6.8

The Major Difficulties Faced by Selected Households during the Preceding Year

(Percentage of households)

	Marginal	*Small*	*Medium*	*Large*	*Landless*	*Total*
Punjab						
Health related problems	37.66	39.02	44.83	29.41	45.16	39.44
Lack of money or poverty	5.19	10.98	6.90	9.41	7.26	7.98
Shortage of basic facilities	22.08	19.51	18.97	16.47	14.52	17.84
Agricultural related problems	15.58	13.41	15.52	22.35	19.35	17.61
Lack of employment	7.79	2.44	5.17	7.06	7.26	6.10
Litigation related problems	11.69	14.63	8.62	15.29	6.45	11.03
Bihar						
Health related problems	58.62	58.62	56.45	44.83	47.30	55.40
Lack of money or poverty	16.67	14.94	14.52	24.14	21.62	17.37
Shortage of basic facilities	13.22	10.34	12.90	6.90	12.16	11.97
Agricultural related problems	1.72	6.90	6.45	10.34	2.70	4.23
Lack of employment	2.30	2.30	4.84	3.45	12.16	4.46
Litigation related problems	4.02	4.60	4.84	10.34	1.35	4.23
Others	3.45	2.30	0.00	0.00	2.70	2.35

Source: Field Survey.

To the question of what was the biggest thing households were lacking, around 35 per cent of the households in Punjab and 40 per cent in Bihar reported lack of the basic facilities like connecting roads, electricity, good schools and hospitals. Shortage of money at times was another irksome problem reported by 30 to 35 per cent of the households in the two states. Employment opportunities for the educated youth and other supplementary employment opportunities were lacking in the peripheral areas according to the respondents in both the states (Table 6.9).

Table 6.9

Problems Identified as most Important by the Respondents

(Percentage of households)

	Marginal	*Small*	*Medium*	*Large*	*Landless*	*Total*
Punjab						
Basic facility	32.47	25.61	27.59	42.35	43.55	35.68
Agricultural related	19.48	10.98	18.97	16.47	4.03	12.68
Shortage of money	27.27	51.22	36.21	28.24	38.71	36.62
Lack of employment	9.09	8.54	5.17	5.88	4.03	6.34
Shortage of goods	9.09	3.66	10.34	4.71	8.87	7.28
Others	2.60	0.00	1.72	2.35	0.81	1.41
Bihar						
Basic facility	41.95	39.08	30.65	27.59	50.00	40.14
Agricultural related	12.07	13.79	37.10	37.93	6.76	16.90
Shortage of money	33.91	32.18	19.35	34.48	28.38	30.52
Lack of employment	8.05	8.05	6.45	0.00	5.41	6.81
Shortage of goods	2.87	4.60	1.61	0.00	2.70	2.82
Others	1.15	2.30	4.84	0.00	6.76	2.82

Source: Field Survey.

In the case of lack of infrastructure at the community or village level, a majority of the households in Bihar complained about the lack of physical infrastructure like roads, transportation, communication, electricity etc., while in Punjab it was the problem of social infrastructure like schools, community centres, health centres etc., which has become a stumbling block for the selected households. However, 24 per cent of the households in Bihar also reported lack of social infrastructure as their basic problem. Similarly, 22 per cent of the households in Punjab reported physical infrastructure as their major problem. A very small number of respondents, less than 5 per cent, quoted financial and market infrastructure as their major problem in both the states (Table 6.10).

About 36 per cent of the respondents in Punjab and 41 per cent in Bihar were of the view that their plight could change by having sufficient acreage of land to till (Table 6.11). This was the view of not only marginal and landless landholders but also of the medium and large farmers. Around 50 per cent of the respondents in the category of small and medium

Table 6.10

Problems of Infrastructure in the Villages

(Percentage of households)

	Marginal	*Small*	*Medium*	*Large*	*Landless*	*Total*
Punjab						
Physical infrastructure	15.58	21.95	31.03	24.71	18.55	21.60
Social infrastructure	59.74	58.54	55.17	63.53	73.39	63.62
Financial infrastructure	6.49	2.44	1.72	4.71	0.00	2.82
Village social infrastructure	2.60	0.00	1.72	2.35	2.42	1.88
Marketing infrastructure	3.90	0.00	3.45	1.18	0.00	1.41
Others	11.69	17.07	6.90	3.53	5.65	8.69
Bihar						
Physical infrastructure	51.15	50.57	43.55	48.28	48.65	49.30
Social infrastructure	24.14	19.54	22.58	34.48	25.68	23.94
Financial infrastructure	1.15	3.45	3.23	3.45	0.00	1.88
Village social infrastructure	1.15	3.45	4.84	4.84	8.11	3.29
Marketing infrastructure	0.00	0.00	0.00	0.00	1.35	0.23
Others	22.41	22.99	25.81	13.79	16.22	21.36

Source: Field Survey.

farmers and around 25 per cent among the large farmers expressed that cultivating higher area could ameliorate their major sufferings. However, almost the same number of respondents in Punjab and Bihar wanted to opt out of farming and go for service, business or trading activities if the opportunity arose, to get a solution to their key problems. Noticeably, very few respondents, only 10 per cent, viewed the government as a messiah who would deliver them from their suffering.

Changes in Income and Input Availability

In our bid to understand the respondents' views on changing government policies and their impact on the rural lifestyle, we asked few direct questions related to their income, farming assets and other implements and their expectations from the government. On the question whether respondents' income has gone up or down during the last 10-year period since the globalisation process started in the agriculture sector, around 41 per cent of the households in Punjab and 58 per cent in Bihar, were of the opinion that their income has gone up during this period. Around 25 per cent of the respondents in Punjab and 18 per cent in Bihar felt the opposite, while 34 per cent in Punjab and 24 per cent in Bihar were of the opinion that there was no significant difference in their income during the last 10 years.

Table 6.11

Respondents' Suggestion for Ameliorating their Problems

(Percentage of households)

	Marginal	*Small*	*Medium*	*Large*	*Landless*	*Total*
Punjab						
Intoxicants prohibition	9.09	2.44	3.45	5.88	2.42	4.46
Raising service/business opportunities	31.17	26.83	24.14	47.06	43.55	36.15
Help from government	7.79	15.85	10.34	12.94	11.29	11.74
Sufficient agricultural land	36.36	43.90	50.00	21.18	33.87	35.92
Others	15.58	10.98	12.07	12.94	8.87	11.74
Bihar						
Intoxicants prohibition	4.60	6.90	12.90	10.34	2.70	6.34
Raising service/business opportunities	29.31	14.94	17.74	31.03	22.97	23.71
Help from government	12.07	4.60	6.45	17.24	9.46	9.62
Sufficient agricultural land	34.48	58.62	38.71	24.14	41.89	40.61
Others	19.54	14.94	24.19	17.24	22.97	19.72

Source: Field Survey.

A majority of the households who felt strongly that their income has risen during this period were large and medium farmers in both the states. In Punjab, around 50 per cent of the large and medium farmers and in Bihar around 70 to 80 per cent of these farmers felt that their incomes had improved. Among marginal farmers and landless labourers, only 20 to 30 per cent in Punjab and 40 to 50 per cent in Bihar opined that their income has increased. On the other hand, above 30 to 40 per cent of the marginal and landless farmers in Punjab and around 25 per cent in Bihar were disappointed by the policies of the last 10 years, which led to fall in their income during this period. No large farmers and only 5 per cent medium farmers in Bihar and around 15 per cent of the large and medium farmers in Punjab expressed the view that their income fell during this period. Thus, we can conclude that although a majority of farmers of all sizes had a positive opinion about the ongoing liberalisation, large farmers appeared to be at a more advantageous position than their counterpart small and marginal farmers.

Table 6.12

Respondents' View on Change in Household Income since 1995

(Percentage of households)

	Marginal	*Small*	*Medium*	*Large*	*Landless*	*Total*
Punjab						
Punjab						
Gone up	37.66	52.44	51.72	50.59	24.19	41.08
Down	31.17	23.17	13.79	16.47	34.68	25.35
Stayed about the same	31.17	24.39	34.48	32.94	41.13	33.57
Bihar						
Gone up	51.19	65.88	71.67	82.76	43.84	58.07
Down	23.81	18.82	5.00	0.00	23.29	18.31
Stayed about the same	25.00	15.29	23.33	17.24	32.88	23.61

Source: Field Survey.

The respondents were of the opinion that supplementary income, (which they earned either through subsidiary activity along with the main farming activity or through permanent employment of some family members in these activities), was very important and it provided them the most needed support at the time of scarcity. Around 37 per cent of the households in Punjab and 32 per cent in Bihar viewed agriculture as the main activity that made them better off than their other companions (Table 6.13). However, 38 per cent of the households in Bihar were of the view that earnings from a regular job in the government or private sector provided them the much needed cushion, and not agricultural earnings. Similar views were expressed by Punjab farmers for the earnings they were getting from side businesses like village shops or retail shops in the nearby town, *arhatia* or commission agent shops in the *mandi*, small mills in the village etc.

Table 6.13

Factors Identified as Most Important for Making the Household Relatively Better Off

(Percentage of households)

	Marginal	*Small*	*Medium*	*Large*	*Landless*	*Total*
Punjab						
Agri. land and implements	27.27	45.12	31.03	31.76	44.35	37.09
Income from business	38.96	21.95	39.66	38.82	30.65	33.33
Job/service	19.48	14.63	15.52	10.59	11.29	13.85
Hard work	6.49	7.32	5.17	5.88	7.26	6.57
Others	7.79	10.98	8.62	12.94	6.45	9.15
Bihar						
Agri. land and implements	35.06	18.39	32.26	27.59	44.59	32.39
Income from business	8.62	13.79	16.13	6.90	2.70	9.62
Job/service	37.93	47.13	32.26	41.38	36.49	38.97
Hard work	16.67	20.69	19.35	24.14	14.86	18.08
Others	1.72	0.00	0.00	0.00	1.35	0.94

Source: Field Survey.

The successful breakthrough in productivity in Indian agriculture was the result of high yielding variety seeds coupled with fertilisers, pesticides and irrigation. In order to check whether the access to these inputs has increased or declined after the globalisation process has taken over, we asked this question directly to the respondents. Almost all the respondents witnessed some marginal increase in the access to these inputs during the last ten years in both the states. Around 90 per cent of the households belonging to all categories in Punjab opined that they had access (though possibly not sufficient for their requirements) to institutional credit, chemical fertilisers, pesticides and HYV seeds. Moreover, access to these inputs somewhat increased during the reforms period. In Bihar, 73 per cent of the households denied any access to institutional credit as was also seen in the agricultural credit section. They were completely dependent on informal sources for the credit. Access to other inputs in Bihar was at par with that in Punjab. As in Punjab, respondents in Bihar also felt that their access to these inputs increased during the reform period. To the question of institutional credit in Bihar, 55 per cent of the large farmers confirmed their access to such credit. In the case of marginal and small farmers, the percentage of households who accepted access to such credit was only 20 and 30 per cent, respectively.

Table 6.14

Respondents' Access to Agricultural Inputs: Pre- and Post-Reforms

(Percentage of households)

		Marginal	*Small*	*Medium*	*Large*	*Total*
Punjab						
Institutional credit	1995	87.32	92.41	87.50	87.65	88.85
	Current	88.73	91.14	89.29	88.89	89.55
Chemical fertiliser	1995	91.55	91.14	94.64	93.83	92.68
	Current	95.77	94.94	92.86	95.06	94.77
Pesticides	1995	85.92	88.61	89.29	92.59	89.20
	Current	88.73	91.14	92.86	93.75	91.61
HYV Seeds	1995	74.65	77.22	82.14	82.72	79.09
	Current	81.69	81.01	78.57	83.95	81.53
Bihar						
Institutional credit	1995	21.05	29.89	25.81	55.17	26.93
	Current	19.88	32.18	25.81	51.72	26.65
Chemical fertiliser	1995	96.47	91.95	98.39	92.86	95.39
	Current	99.42	94.25	98.39	93.10	97.42
Pesticides	1995	80.12	86.21	91.94	72.41	83.09
	Current	90.59	93.10	95.16	89.66	91.95
HYV Seeds	1995	77.19	80.46	87.10	79.31	79.94
	Current	92.40	87.36	91.94	86.21	90.54

Source: Field Survey.

Although the majority of the farmers admitted to access to agriculture inputs, availability of these inputs in the desired quantity was a moot question. We attempted to find out whether there was any inadequacy in their availability and if so, what were the reasons thereof. The responses of the selected households are summarised in Table 6.15. In the respondents' view the adequacy of the above discussed inputs improved tremendously in Punjab during the last decade although there were other problems like high prices of these inputs, no provision from the government and some of these inputs being out of reach for some respondents in the wake of their low purchasing power. In Bihar, however, around 80 per cent of the respondents expressed access to institutional credit and chemical fertilisers as either

Table 6.15

Reasons for Inadequacy in the Availability of Agricultural Inputs

(Percentage of household)

		Marginal		*Small*		*Medium*		*Large*		*Total*	
		1995	*Current*	*1995*	*Current*	*1995*	*Current*	*1995*	*Current*	*1995*	*Current*
Punjab											
Institutional credit	Inadequate at the moment	25.00	14.29	83.33	42.86	57.14	33.33	50.00	33.33	51.61	31.04
	Out of reach	50.00	28.57	16.67	14.29	0.00	0.00	10.00	22.22	19.35	17.24
	Do not get from the govt.	25.00	57.14	0.00	42.86	42.86	66.67	40.00	44.44	29.03	51.73
Chemical fertiliser	Inadequate at the moment	40.00	0.00	16.67	33.33	66.67	0.00	40.00	0.00	36.84	8.33
	Out of reach	20.00	0.00	0.00	0.00	0.00	33.33	0.00	0.00	5.26	8.33
	Prices are high	0.00	0.00	16.67	33.33	0.00	0.00	20.00	25.00	10.53	16.67
	Do not get from the govt.	40.00	100.0	50.00	33.33	0.00	66.67	40.00	75.00	36.84	66.67
	Others	0.00	0.00	16.67	0.00	33.33	0.00	0.00	0.00	10.53	0.00
Pesticides	Inadequate at the moment	40.00	0.00	37.50	14.29	50.00	0.00	0.00	0.00	33.33	4.17
	Prices are high	0.00	25.00	0.00	0.00	0.00	25.00	16.67	20.00	3.33	16.67
	Not available in the fair shop	40.00	37.50	37.50	57.14	50.00	50.00	50.00	40.00	43.33	45.83
	Do not get from the govt.	10.00	37.50	25.00	28.57	0.00	25.00	33.33	40.00	16.67	33.33
	Others	10.00	0.00	0.00	0.00	0.00	0.00	0.00	0.00	3.33	0.00
HYV seeds	Inadequate at the moment	11.76	0.00	22.22	6.67	40.00	16.67	21.43	7.69	22.03	7.55
	Prices are high	0.00	7.69	0.00	0.00	0.00	8.33	7.14	7.69	1.69	5.66
	Not available in the fair shop	47.06	30.77	44.44	46.67	0.00	0.00	35.71	30.77	35.59	28.30
	Do not get from the govt.	35.29	61.54	33.33	46.67	60.00	75.00	35.71	53.85	38.98	58.49
	Others	5.88	0.00	0.00	0.00	0.00	0.00	0.00	0.00	1.69	0.00

contd...

...contd...

		Marginal		Small		Medium		Large		Total	
		1995	Current	1995	Current	1995	Current	1995	Current	1995	Current
Bihar											
Institutional credit	Inadequate at the moment	19.55	20.00	13.55	12.28	39.13	41.30	38.46	50.00	22.71	23.81
	Out of reach	69.17	66.67	69.49	68.42	39.13	36.96	61.54	42.86	63.35	60.32
	Do not get from the govt.	11.28	13.33	16.95	19.30	21.74	21.74	0.00	7.14	13.94	15.87
Chemical fertiliser	Inadequate at the moment	100.0	80.00	71.43	66.67	100.0	100.0	100.0	100.0	86.67	77.78
	Prices are high	0.00	20.00	14.29	16.67	0.00	0.00	0.00	0.00	6.67	11.11
	Do not get from the govt.	0.00	0.00	14.29	16.67	0.00	0.00	0.00	0.00	6.67	11.11
Pesticides	Inadequate at the moment	53.33	6.67	53.85	16.67	20.00	0.00	50.00	66.67	50.00	14.81
	Prices are high	6.67	13.33	15.38	16.67	0.00	0.00	37.50	33.33	12.50	14.81
	Not available in the fair shop	23.33	33.33	23.08	66.67	80.00	100.0	12.50	0.00	26.79	44.44
	Not in nearby place	3.33	6.67	7.69	0.00	0.00	0.00	0.00	0.00	3.57	0.00
	Do not get from the govt.	6.67	40.00	0.00	0.00	0.00	0.00	0.00	0.00	3.57	22.22
	Others	6.67	0.00	0.00	0.00	0.00	0.00	0.00	0.00	3.57	3.70
HYV seeds	Inadequate at the moment	33.33	23.08	52.94	63.64	50.00	40.00	50.00	50.00	41.43	42.42
	Out of reach	5.13	23.08	0.00	9.09	12.50	20.00	0.00	0.00	4.29	15.15
	Prices are high	35.90	53.85	35.29	18.18	37.50	40.00	50.00	25.00	37.14	36.36
	Not available in the fair shop	12.82	0.00	5.88	0.00	0.00	0.00	0.00	25.00	8.57	3.03
	Not in nearby place	5.13	0.00	0.00	0.00	0.00	0.00	0.00	0.00	2.86	0.00
	Do not get from the govt.	0.00	0.00	5.88	9.09	0.00	0.00	0.00	0.00	1.43	3.03
	Others	7.69	0.00	0.00	0.00	0.00	0.00	0.00	0.00	4.29	0.00

Source: Field Survey.

inadequate at the moment or out of their reach. Moreover, this scenario remains as grim as it was a decade earlier. In the case of pesticides and HYV seeds, although they are now adequately available, yet a majority of the households complained that their open market prices are very high and they were hardly available at controlled rates at fair price shops.

The major hurdles faced by the farmers in their agricultural operations were the lack of quality inputs, improper price policy in the agriculture sector and the lack of agricultural infrastructure. Almost half of the selected households in Punjab and 70 per cent in Bihar had reservations about the available infrastructure in the agriculture sector, e.g., lack of proper banks and credit facilities, lack of markets or regulated *mandis*, *kachha* roads and improper means of transportation etc. In Punjab, a majority of the households were of the view that the price policy of the government was not favourable to the agricultural sector. In their opinion, the input prices, i.e., the prices of fertilisers, pesticides, tractors and implements have risen at a very rapid pace during the last decade whereas the price of various crops produced by them was increasing at a snail's pace. Their specific problem was the very high diesel prices that had made agriculture uneconomic, as during last 10 to 15 years, most of the farmers had shifted from manual agriculture dependent on bullocks to mechanical farming,

Table 6.16

Main Difficulties Faced in Production Process

(Percentage of households)

	Marginal	*Small*	*Medium*	*Large*	*Total*
Punjab					
Agri. input quality and price	27.63	36.59	46.55	50.59	40.20
Agri. infrastructure	55.26	57.32	37.93	44.71	49.50
Inadequate governance	14.47	3.66	15.52	4.71	8.97
Others	2.63	2.44	0.00	0.00	1.33
Bihar					
Agri. input quality and price	29.82	22.09	27.42	20.69	26.72
Agri. infrastructure	66.08	75.58	67.74	75.86	69.54
Inadequate governance	4.09	2.33	4.84	3.45	3.74

Source: Field Survey.

whereby every farming operation depended on the use of the tractor. The unfair trade in spurious pesticides and insecticides was another irksome problem faced by the respondents in both the states. Inadequate infrastructure for procurement, lack of research and extension and un-preparedness to deal with unforeseen events, like droughts and floods, was also highlighted by the respondents.

Collective Action by the Villagers

In reply to our question what villagers could do collectively for the welfare of the village, a majority viewed improving social infrastructure as the prime concern of the villagers. One of the most important activities among the social infrastructure they wanted to develop was a school and hospital/dispensary for the village (Table 6.17). Building a *pucca* or metalled road was also given priority by a large number of respondents. Ironically, developing an agricultural market or water storage tanks received the least priority by the respondents in both the states.

Table 6.17

Priorities for Collective Action for Village Development

(Percentage of households)

	Marginal	*Small*	*Medium*	*Large*	*Landless*	*Total*
Punjab						
Pucca street/*sadak*	16.88	14.63	18.97	12.94	16.94	15.96
Market	7.79	7.32	8.62	5.88	6.45	7.04
Water storage	0.00	2.44	1.72	1.18	0.81	1.17
School	31.17	20.73	12.07	34.12	16.13	22.77
Hospital/dispensary	24.68	21.95	15.52	17.65	29.03	22.77
Others	19.48	32.93	43.10	28.24	30.65	30.28
Bihar						
Pucca street/*sadak*	29.31	26.44	35.48	20.69	41.89	31.22
Market	16.67	12.64	4.84	24.14	10.81	13.62
Anything else	21.84	33.33	25.81	10.34	14.86	22.77
School	6.90	10.34	9.68	10.34	22.97	11.03
Hospital/dispensary	25.29	17.24	24.19	34.48	9.46	21.36

Source: Field Survey.

Table 6.18

Major Obstacles in Collective Action

(Percentage of households)

	Marginal	Small	Medium	Large	Landless	Total
Punjab						
Divided in two groups	32.47	23.17	29.31	30.59	23.39	27.23
Lack of cooperation	24.68	18.29	20.69	21.18	17.74	20.19
Jealousy about others	7.79	14.63	15.52	12.94	13.71	12.91
Lack of govt. help	11.69	17.07	13.79	9.41	20.97	15.26
Lack of proper guidance	16.88	10.98	12.07	21.18	18.55	16.43
Lack of awareness	6.49	15.85	8.62	4.71	5.65	7.98
Bihar						
Divided in two groups	9.20	5.75	3.23	3.45	10.81	7.51
Lack of cooperation	47.13	49.43	43.55	51.72	58.11	49.30
Jealousy about others	2.87	3.45	4.84	0.00	9.46	4.23
Lack of govt. help	13.22	8.05	11.29	6.90	4.05	9.86
Lack of proper guidance	12.64	11.49	14.52	20.69	6.76	12.21
Lack of awareness	14.94	21.84	22.58	17.24	10.81	16.90

Source: Field Survey.

To our question why villagers were not acting collectively for the welfare of the village, if it was possible to do so, a number of reasons were cited by the respondents. Groupism, lack of cooperation among themselves and lack of proper guidance were the prime reasons given for this. Group politics was the prime reason quoted in Punjab while lack of cooperation was given as the biggest reason by the respondents in Bihar. Lack of equal government help, jealousy towards each other and lack of awareness on the part of villagers were other reasons given by the respondents (Table 6.18).

Further, to a question on what help they desired from the government, 58 per cent of the respondents in Bihar sought an improvement in the physical infrastructure in their surrounding areas. In Punjab, a majority of the households wanted better access to agriculture inputs at cheaper prices. Improvement in financial and social infrastructure was also demanded by a large number of households in Punjab. In Bihar access to cheap and quality inputs was the second most important priority among the selected respondents.

Table 6.19

Areas for State Intervention

(Percentage of households)

	Marginal	*Small*	*Medium*	*Large*	*Total*
Punjab					
Input availability and cost	30.26	37.80	39.66	43.53	37.87
Physical infrastructure	14.47	17.07	8.62	14.12	13.95
Social infrastructure	28.95	15.85	18.97	9.41	17.94
Financial Infrastructure	19.74	21.95	18.97	17.65	19.60
Marketing infrastructure	3.95	7.32	12.07	14.12	9.30
Others	2.63	0.00	1.72	1.18	1.33
Bihar					
Input availability and cost	27.65	29.07	24.19	27.59	27.38
Physical infrastructure	56.47	53.49	67.74	55.17	57.64
Social infrastructure	4.12	5.81	0.00	6.90	4.03
Financial Infrastructure	6.47	3.49	6.45	10.34	6.05
Marketing infrastructure	4.71	8.14	1.61	0.00	4.61
Others	0.59	0.00	0.00	0.00	0.29

Source: Field Survey.

7 Lessons for Policy

Economic Reforms and Agriculture Sector

The liberalisation and globalisation strategy followed by India in the early nineties began with a few policy changes. These included: scrapping of the industrial licensing regime, reduction in the number of areas reserved for the public sector, amendment of the Monopolies and Restrictive Trade Practices Act (1969), start of the privatisation programmes, reduction in tariff rates, and change over to market determined exchange rates. Over the years there has been a steady liberalisation of the current account transactions, more and more sectors have been opened up for foreign direct investments and portfolio investments facilitating entry of foreign investors in telecom, roads, ports, airports, insurance and other major sectors.

The focus of the reform process was initially on the manufacturing sector, but gradually these changes were extended to the other sectors of the economy including agriculture. In agriculture, although there is no direct government intervention in production and investment decisions of farmers, the government influences the legal and economic environment in which farmers and other economic agents operate. In the case of the domestic market, the withdrawal of restrictions on the movement of agricultural commodities was the major step initiated by the reform process in agriculture. A number of changes were also made in the Essential Commodities Act (1955), such as the removal of the licensing requirements and stocking limits for the wholesale and retail trade. Selective credit controls that were used to regulate institutional credit to traders have also been abandoned. State trading activities, once the bastion of full government control over agricultural trade, have been curtailed in almost all products. Future markets earlier banned under various statutory orders are also now allowed in food grains and a number of other commodities. In the case of external trade, barring a few exceptions, the exports of all major agricultural commodities have been liberalised. Market access has been extended to agricultural products by relaxing licensing arrangements, reducing tariffs, freeing more items from quantitative restrictions, and allowing the private sector to import food items. The private sector may import oilseeds and edible oils, and private millers may import wheat and corn in certain conditions.

The tariff rates have been reduced sharply over the decade from a weighted average of 72.5 per cent in 1991-92 to 24.6 per cent in 1996-1997. Though tariff rates went up slowly in the late nineties, they touched 35.1 per cent in 2001-02. In another two years most non-tariff barriers were dismantled by March 2002, including almost all quantitative restrictions. The growth rates generated in manufacturing, especially in the service sector, during the process of reforms are creating pressure for change in agricultural policy. Food demand is expanding and diversifying, which, in turn, is placing new demands on market infrastructure and institutions. Long-standing production procurement, and distribution programmes that focus on cereals are becoming increasingly expensive and out of step with market demand. At the same time, rising subsidy outlays appear to be constraining public investment in transforming institutions and market infrastructure. Finally, despite improved growth in the overall economy, growth in farm output and, particularly, rural employment, is slowing down.

National elections in 2004 brought to the fore the dismal performance in the agriculture sector and the necessity for reforms, but policymakers are still seeking a consensus on specific and fundamental reforms. Food consumption patterns, in India have diversified significantly. Consumption of fruits, vegetables, edible oils, and animal products is rising much faster than that of wheat and rice, the staple grains in the Indian diet. Higher incomes, particularly in lower and middle-income households, are having an important impact on food demand in India, because these groups tend to spend a relatively large share of their income on food consumption. Middle-income and urban consumers are likely to spend more of their income on upgrading and diversifying their diets. Eating out more often and eating processed and convenience foods is becoming a practice now.

Indian producers are responding to rising demand with only partial success. Recent trade liberalisation measures have introduced new products at lower prices, thus creating competitive pressures for domestic producers. Constraints such as poor infrastructure, inefficient markets, and low investment also restrict Indian producers' ability to satisfy consumer demand. Consumer demand for greater variety, coupled with more liberal import policies, is pressuring India's producers and marketing system to provide a broader range of products at competitive prices. However, Indian agriculture is characterised by low productivity, with average crop yields well below world levels. Large investments, public and private, are needed to improve seed varieties and irrigation and plant protection practices. Government agencies are promoting diversification in production, research, and farm extension, but successful diversification is likely to require shifting public resources away from subsidies and improving incentives for private investment.

By now, India has about a decade's experience of living under the 'global umbrella'. The problems and stresses of switching over to an open economy are getting familiar. That a sort of nationwide political consensus on the need for economic reforms exists, is a redeeming feature of the Indian polity. What the effects of the reforms implemented so far on various aspects of people's living are, what directions the economy is likely to take, and how people's working, living and general standard of living would look like after a decade or so, are the questions that are yet to be answered with empirical certainty. This study is an attempt in this very direction.

Our study looked at the impact of the agriculture related policy changes on the emerging production, marketing, employment and earning status of small *versus* large farmers. The findings of the study are summarised below.

Summary of the Findings of the Study

The comparative analysis of Punjab and Bihar agriculture presents quite a contrasting picture. Whereas Punjab has remained the granary of India since the Green Revolution in the sixties and seventies, Bihar lagged far behind in the adoption of new technology, productivity and rural infrastructure. While agriculture remains the prime activity and provides employment to more than two-thirds of the population in both the states, the importance of agriculture in gross production is falling at a faster rate in Punjab compared to Bihar.

Further, the rate of transformation of Bihar's economy in terms of growing importance of non-farm and non-agricultural sector was much slower compared to Punjab and other states. Lack of industrialisation and poor rural-urban growth linkages have further worsened the employment position in Bihar, so the percentage of population dependent on agriculture or disguisedly underemployed in agriculture was relatively higher in Bihar compared to Punjab.

Land Use

A large share of cultivable area still remains unused in Bihar, whereas almost all the cultivable land is under the plough in Punjab. Vertical intensification was high in both the states, as cropping intensity was measured at 1.4 in Bihar, which compares well with 1.9 in Punjab, given the fact that 95 per cent of gross cropped area was irrigated in Punjab as against 48 per cent in Bihar. Around 90 per cent of holdings in Bihar were marginal in contrast to only 13 per cent in the case of Punjab. Over time, holdings were getting further fragmented in Bihar in keeping with the trend prevalent in most of the states. In Punjab, however, a different trend of consolidation of holdings was visible, thanks to the prevalence of the phenomenon of 'reverse tenancy': Because of high commercialisation and mechanisation of agriculture and due to increasing cost of cultivation, the marginal holdings in Punjab were becoming unviable therefore, marginal farmers were switching out of agri-business in favour of either wage earnings or other formal or informal employment.

All households barring the category of landless in Punjab leased in more land than they leased out, but the large farmers leased-in much higher area than their counterpart small farmers. Many of the small farmers were leasing out land in favour of medium and large farmers. In Bihar on the other hand, it was the marginal and small farmers who leased in proportionately higher area compared to other categories, while large farmers were net losers as they were leasing out more land than leasing in. Unlike Punjab where tenancy was mostly in monetary terms, in Bihar a very high proportion of tenancy was still in barter terms, either in kind or in sharecropping.

Land Ownership

The pattern of land distribution among the selected cultivators presented the broad trends prevailing in these two states. The average size of holdings among the selected households was 6.7 acres in Punjab that was much higher compared to 3.4 acres in Bihar. More than 75 per cent of the holdings and 68 per cent of the area was less than 10 acres in Bihar, while in Punjab this ratio was 50 and 32 per cent, respectively. Whereas in Bihar the large holdings were less than 7 per cent occupying only 32 per cent of the area, the number of large holdings in Punjab was 20 per cent occupying 68 per cent of the operated area.

Living Standards

On an average, the lifestyle reflected by consumer and productive assets in Punjab was far better than that in Bihar. In Punjab around 95 per cent of the households had *pucca* or mixed houses made of bricks and only 5 per cent of the households were staying in mud houses. In Bihar, 42 per cent of the households were staying in mud houses and another 19 per cent had mixed houses made of bricks and mud. Likewise, 88 per cent of the houses in Punjab had domestic electricity connections, whereas in Bihar only 27 per cent of the households had domestic electricity connections. While in Punjab around 60 to 70 per cent of the cultivators had electric tube well connections, in Bihar less than 10 per cent of the large farmers and less than 1 per cent of the medium, small and marginal farmers had electric tube well connections.

Ownership of Livestock

Almost all categories of households in both the states were rearing buffaloes/cows for milk purposes. The percentage of households keeping milch animals had a direct relationship with farm size. The households keeping more than six such animals most likely for commercial purposes were less than 5 per cent in Punjab and less than 1 per cent in Bihar. Among our Census households in both the states, rearing milch animals was the only major subsidiary activity undertaken by the farmers either for supplementing income through selling milk or rearing buffaloes and cows for their personal consumption of milk.

Crop Diversification

The most desired crop diversification in both the states was hardly followed by the households as those growing high value cash crops like fruits, vegetables and flowers were less than 5 per cent. On an average, only 6 per cent of the households in Punjab and 4 per cent in Bihar were able to diversify their income in supplementary activities other than the main crops and dairy farming, which remained the major source of employment as well as earnings of the rural households. Among all the households, a small number of mostly medium and large farmers in Punjab had been engaged in contract farming to grow potato and *basmati* paddy with Pepsi and Hindustan Lever. However, in both the states no farmer was found growing any crop for export purposes or selling his produce overseas through any intermediaries.

Pattern of Employment

Following the National Classification of Occupations (1968), it was seen that around 35 to 40 per cent of the population was the truly active population among the sample households who were engaged in some earning activities either in agriculture and related activities, or in the secondary and tertiary sectors.

As regards the employment pattern in Punjab and Bihar, around 45 to 50 per cent of the households in both these states were directly employed in agriculture and related activities. Non-agricultural labour in the secondary and tertiary activities, public and private sector jobs and self-employment in business and crafts were the other minor sources of employment for the rural households. Among the landless households, only 30 per cent were employed in agriculture and the remaining 70 per cent depended on non-agricultural activities.

The trends in occupation showed that out of 35 per cent working members in both the states, around 23 and 21 per cent were engaged in agriculture as cultivators or labourers and other allied activities like plantation, dairy etc., in Punjab and Bihar respectively. In Punjab the percentage was highest, 32 per cent for medium farmers, and lowest—around 7 per cent—for landless labourers. In Bihar, the percentage was highest for small farmers, 26 per cent, and lowest, 11 per cent, for the landless labourers.

In the crop sector, the maximum employment was in harvesting activities, followed by caring, sowing and marketing activities respectively in that order. The total labour absorption in harvesting, sowing and other activities in Punjab were calculated as 40 man-days per acre. The number of man-days declined from 50 in the case of marginal farmers to 43 on small farms and further to 35 in the case of medium farmers and again increased to 41 in the case of large farmers. In Bihar, on an average 43 man-days per acre were employed in the harvesting and sowing activities, and there was a clear negative relation between labour employment and increasing farm size. The labour absorption was above 65 man-days per acre on marginal farms, which declined to 32 man-days in the case of large farms. Use of family labour showed an inverse relationship with farm size in both the states. The ratio of hired labour to family labour was much higher in Punjab as compared to Bihar.

Among the subsidiary activities, livestock and dairy farming was the leading one. Almost all households kept milch animals for household milk consumption, and in some cases for commercial purposes also. Farming was the primary activity for cultivators in Punjab while in Bihar a significant number of cultivators had farming as their secondary occupation, possibly because of low returns from agriculture in the state. The remaining 10 to 15 per cent of the working population not engaged in agriculture and allied activities were engaged in manufacturing/service sectors as government employees (regular or ad hoc basis), transport operators or as construction workers.

Among the agriculture-allied activities, dairy farming or livestock rearing was the prominent one among the selected households in both the states. In Punjab, family labour contributed 70 per cent of its time to livestock activities and only 30 per cent was devoted to crop sector activities. Similarly in Bihar the ratio of livestock and crop activities was 64 and 36 per cent, respectively. Comparing manpower in crop and livestock activities, the data shows that in Punjab, around 59 per cent of the manpower was engaged in crop sector and the remaining 41 per cent found employment in the livestock sector. In Bihar the ratio of crop and livestock employment was 54 and 46 per cent, respectively. Across the farm size categories, small and marginal farmers devoted comparatively more time to livestock rearing, while large farmers allocated more time for the management of crop production.

In non-farm employment, around 4 per cent in Punjab and 6 per cent in Bihar were employed in regular government, semi-government or private sector jobs. Landless labourers had the maximum percentage of members seeking their daily earnings as ad hoc and casual labourers in both the states.

Farm and Non-Farm Earnings

Comparing farm and non-farm earnings, it was observed that the farm income was the chief source of earning in Punjab, which contributed to around 76 per cent of total income for the selected households. In addition to farm income, livestock rearing contributed around 9 per cent earnings, thus making the share of agriculture and allied activities 85 per cent. Among the non-agricultural employment activities, government/ semi-government jobs, casual labour, self-employment and remittances from outside were the other major sources of household earnings. Within farm size classes, non-farm income was more important for marginal and small farmers who earned 50 to 70 per cent from agriculture and related activities, and the rest was contributed by non-farming activities. Large farmers, on the other hand, earned more than 90 per cent of their income from farm and related activities. Livestock rearing was particularly significant for the marginal and small farmers as it contributed around 22 per cent earnings for the marginal farmers and 18 per cent for the small farmers, whereas large farmers had only 7 per cent of their total income from livestock. Casual labour was a major source of earning for the landless labourers although it also supplemented income for the marginal farmers. Self-employment contributed more than 10 per cent earnings in the case of marginal farmers and landless labourers, while in other classes its share was less than 2 per cent. Besides these activities, government and semi-government jobs also contributed around 8 to 9 per cent in the case of marginal, small and medium farmers.

Bihar presents a completely different picture in terms of farm and non-farm earnings. The share of farm income in total household income was only 36 per cent in comparison to 76 per cent in the case of Punjab. The share of livestock activities was also very small, only 4 per cent whereas, in Punjab it was above 8 per cent. Thus farming (and allied activities) in Bihar, although being full time occupation for the cultivators, was not able to provide even half of their earnings, forcing them to largely depend on non-farming activities to make both ends meet. Jobs in the government or private sector, self-employment, old age pensions, casual earnings and outside remittances were the major off-farm employment activities in Bihar. Across farm size categories, per household farm income was not as diverse as it was in Punjab. Share of farm business in total household income was 22 and 38 per cent respectively for marginal and small farmers while it was 47 per cent for large farmers. The share of livestock activities was almost same across all size classes, which averaged around 4 per cent. Regular jobs in the private or government sectors constituted around 15 to 20 per cent share in total household earnings across all size classes. The share of self-employment was also evenly distributed among different size classes while casual labour was the principal means of earning for landless labourers and marginal farmers alone. Comparing gross household income in Punjab and Bihar, the statistics revealed that per household income in Punjab stood at more than twice that of Bihar. However, the difference was much higher, more than 4 times, in the case of farm business income. Apparently, the off-farm income was much more for Bihar cultivators in comparison to Punjab. Moreover, the difference in total household income between the two states was much lower in the lower strata of operated area as compared to higher strata among the selected households in the two states. Per capita income in Punjab was almost twice that of Bihar in the lower strata and more than twice in the upper strata of operational holdings.

Credit

Loans availed by the Punjab farmers were 10 times greater than those availed by the Bihar farmers. The loan needs of large farmers were much higher than those of small farmers. On the other hand, the amount borrowed per acre had an inverse relationship with farm size as the loan amount borrowed by marginal farmers was much higher compared to loans of large farmers in both the states. Another notable point was that institutional loans were also accessible to marginal and small farmers, although its absolute amount was much lower compared to that of the large and medium farmers. Whereas in Punjab, marginal and small farmers had very good access to institutional loans and they competed very well with their counterpart large farmers, the situation in Bihar was not as favourable. In Punjab, 35 to 40 per cent of loan requirement of the marginal and small farmers was fulfilled by the institutional and other government lending programmes, compared to around 60 per cent in the case of large farmers. In Bihar on the other hand, institutional and other government borrowings fulfilled less than 20 to 25 per cent loan requirement of marginal and small farmers compared to above 95 per cent in the case of large farmers.

Irrigation and other Inputs

Among the households surveyed, the net irrigated area was around 99 per cent of the net operated area (NOA) in Punjab while it was a surprisingly very high 93 per cent in Bihar may be due to survey error of counting low land rainfalls water being lifted by diesel pump as irrigated land. In Punjab, electric tube wells were the main source of irrigation, which accounted for 70 per cent of the total operated area. In Bihar, electric tube wells were a minor source, while 88 per cent of the operated area was irrigated by diesel pump sets. Diesel pump sets were the obvious choice in Bihar as many of the rural areas were still to be electrified in the state. Contribution of canals in the irrigated area among the selected households in both the states was very low, only 2 per cent of NOA. Paddy, being a high water intensive crop, was fully irrigated in Punjab, while in Bihar it had 85 per cent area under irrigation. Area under wheat was mostly irrigated in both the states. Oilseeds and fruits and vegetables were also mostly irrigated in both the states. Coarse grains and pulses had higher area un-irrigated.

Like irrigation, Punjab has remained the leading state in the use of high yielding variety HYV seeds also. Among the selected households, more than 90 per cent of the cropped area was cultivated using HYV seeds in Punjab. Bihar lagged far behind in the use of HYV seeds as the state had only 62 per cent area under HYV. Seed vendors and neighbouring farmers were the main source of information to the farmers in both the states. A very small number of farmers depended on more reliable sources of information, namely the agricultural universities in the state or private seed companies for such information. This reflects ignorance on the part of farmers and apathy/lack of interest by the agricultural universities in the states/country involving the farming community in the research activities going beyond the laboratory.

Cropping Pattern

As regards cropping pattern, food crops occupied major area in both the states. Rice-wheat rotation dominated the cropping cycle as the share of these two crops in total gross cropped area was around 70 per cent in both the states. Despite all efforts of the state government, wheat and rice dominated the cropping pattern in Punjab. In fact among the selected households, the area under paddy had gone up from 28 per cent before liberalisation to 33 per cent in the post-liberalisation period. The area under wheat had come down slightly during this period. The reason for the dominance of rice-wheat rotation was the high relative productivity of these two crops compared to other available alternatives, supported by the availability of HYV seeds, subsidised irrigation and the support price policy of the Government of India.

The domination of medium and large landholders in operational holdings in Punjab, high yield rates and development of regulated market has ensured a much larger proportion of production of rice and wheat arriving at the market place. In contrast, in the case of Bihar, the official agencies played very little role in procurement and as a result there was a relative decline in prices of rice and wheat in Bihar compared to Punjab after mid-nineties when the minimum support price (MSP) of paddy and wheat were substantially raised. This shows that even with low yield level the farmers in Bihar were not getting the benefits of increase of minimum support prices.

Among other major crops namely, potato, fodder and cotton, and other minor crops like coarse grains and pulses, there was hardly any change in the area under cultivation during pre- to post-liberalisation periods in Punjab. Comparing the farm size categories, it was observed that marginal farmers devoted comparatively less area to paddy and vegetables than their counterpart large farmers. The small farmers on the other hand, devoted higher area to maize and fodder crops probably because of lack of irrigation and their higher dependence on livestock compared to large farmers.

In the case of Bihar, around 90 per cent of the total cropped area was under food grains. Though, as we have seen, paddy and wheat were the dominant food crops, unlike Punjab where, cotton was the second leading *kharif* crop, maize occupied a principal place among *kharif* crops in Bihar. In the *rabi* season, apart from wheat, *khesari* and *masoor*, the two pulse crops, occupied a significant area of around 12 per cent in the gross cropped area. Gram, mustard, sugarcane and a few vegetables like potato and onion were the other minor crops grown in Bihar. Comparing the pre- and post-liberalisation periods, apparently liberalisation has not touched the rural areas in Bihar, as no major change was visible in the cropping pattern of the state during the last decade. Among farm size categories, marginal and small farmers planted more area under paddy compared to large farmers, a trend that was completely different in Punjab. Apparently, rice being the staple food in Bihar, subsistence concerns dominated the decision process of these small and marginal farmers.

Economics of Production, Cost and Resource Use Efficiency

The average value of output per household in Punjab was almost five times higher than in Bihar. However, the difference was not as high among marginal and small farmers as among the large farmers. The divergence in per household output might be the outcome of differences in operated land. The value of output per acre in Punjab was more than twice than that of Bihar. The difference in productivity between the two states was much less in the case of marginal farmers, Rs 6,792, which increased to Rs 18,207 in the case of large farmers. While looking at the value of output per acre in Punjab, a positive relation between productivity and farm size was clearly evident from the data. However, the trends were completely opposite in the case of Bihar. In the latter case, value of output per acre decreased as the farm size increased, establishing a negative relationship between the two. Thus, the seemingly positive relation between farm size and productivity in Punjab and the negative relationship in Bihar revives the age-old debate on the inverse farm size productivity relationship in the country. In Punjab, due to mechanisation, the trends of economies of scale on large farming have turned the productivity advantage in the favour of large farmers. Bihar agriculture seems to be caught in the web of subsistence farming with a very high percentage of holdings being marginal and small. Therefore, mechanisation and commercialisation of agriculture in the state is still at a primitive stage, tilting the advantage of input intensity in the favour of small farmers and resulting in higher productivity for them.

The cost statistics revealed very interesting results for Punjab and Bihar. Manure, chemical fertilisers and pesticides constituted the major items in the total cost of production in Punjab. Hired/owned tractors and rentals paid for leasing-in land were next in importance. Expenses incurred on human labour, including the imputed value of family labour were the other important component of cost of cultivation. Comparing the percentage of cost across different farm size categories, it was observed that hired tractors and family labour constituted higher costs to the marginal farmers whereas, own tractor expenses and expenses for leased-in land was comparatively higher in the case of large farmers. Distribution of cost across different inputs in Bihar presented a contrasting picture compared to Punjab. Labour (hired plus imputed value of family labour) was the single largest item, which constituted around 50 per cent of total cost of cultivation in Bihar. The components of modern technology namely, fertiliser-pesticides, irrigation and tractors together accounted for around 22 per cent of total cost of cultivation in Bihar, whereas their contribution was around 52 per cent in the case of Punjab. These figures verify our assertion that Punjab agriculture is completely mechanised and has become commercial, whereas Bihar agriculture is still under subsistence orientation, overburdened with family labour, especially in the case of small and marginal farmers who constituted 90 per cent of total holdings in the state.

Production Function

Interpreting the results of production functions in the two states, it was observed that the coefficient of operated area was significant with opposite signs in the two states. Its positive sign in Punjab indicated higher productivity for large farmers, while the negative sign in Bihar emphasised higher efficiency of small farmers in terms of higher land productivity. Fertilisers, mechanisation and human labour were the other significant variables in the two states. The variable for elasticity of mechanisation in

Bihar was almost half that of Punjab indicating lower productivity of machines in Bihar because of predominance of small farming in the state. On the other hand, the elasticity of human labour in Bihar was almost one and a half times higher than in Punjab indicating higher productivity for manual agriculture in that state. In the case of irrigation, submersible pumpsets contributed significantly to land productivity, while investment in non-submersible tube wells failed to raise the land productivity in Punjab. In Bihar, however, only diesel pump sets were used which also remained insignificant in raising land productivity.

Market Surplus and Price Analysis

Punjab farmers were highly commercialised producing a very high proportion, more than four-fifth of their output, for the market. Bihar farmers on the other hand were highly subsistence farmers who produced more than half of their output for self-consumption. In Punjab, marginal farmers with 75 per cent marketed surplus competed well with large farmers who contributed 84 per cent of their output to the market. In Bihar, on the other hand, the contribution of marginal farmers was only 31 per cent whereas large farmers contributed around 53 of their output to the market. Crop-wise, 77 per cent of the output of wheat was marketed in Punjab compared to 51 per cent in Bihar. On the other hand, rice being the staple cereal in Bihar, only 15 per cent of the output was marketed there compared to 98 per cent of the output marketed in Punjab. Among other major crops, 65 per cent of the production of potato and 95 per cent of cotton was marketed in Punjab. In Bihar, 66 per cent of maize, 50 per cent of *masoor* and 61 per cent of *khesari* was marketed.

Analysis of price revealed that realised prices of all crops except vegetables were higher in Punjab compared to Bihar. The realised price for wheat in Punjab was almost equal to the MSP during the survey year, while the paddy price was above the MSP. On the other hand, the realised prices for wheat and paddy in Bihar were much lower than the MSP announced by the government for these two crops. The principal reason for prices prevailing below to MSP in the case of Bihar seems to be complete absence of procurement operations in the state. The results of the aggregate price index ruled out any discrimination against the marginal or small farmers in both the states. Thus, although the marginal farmer in Punjab was being squeezed by the rising cost of cultivation and stagnating yield, forcing him to quit farming by leasing out land in favour of large farmers, the findings clearly rule out any discrimination on account of the sale of his surplus produce in the market. On the other hand, Bihar lacked in basic market infrastructure in terms of regulated *mandis* and thereby 90 per cent of the output was being disposed off in the village to various intermediaries. There seems to be interlocking of credit and product market operating in the field. Due to lack of market infrastructure, there were hardly any government procurement operations taking place in the state thereby leaving the farmer at the mercy of private trade, which was full of exploitation, with the end result that the realised prices were far lower than the MSPs.

The regression results revealed that the value of output was the most important variable determining the value of marketed surplus. The value of the coefficient was more than one in both the states, indicating that any rise in output led to a more than proportionate increase in marketed surplus. Household size was the other important variable significant at one per cent with a negative coefficient, indicating that an additional member in the family reduced marketed surplus, by raising the retention requirement by 16 per cent in Punjab and 25 per cent in Bihar. In the case of net operated area the results revealed that a one per cent increase in net operated area raised the marketed surplus by 1.3 per cent in Punjab and 1.1 per cent in Bihar. As in the case of aggregate marketed surplus, output of each crop was also the most significant variable affecting the marketed surplus of the respective crops in both the states. The magnitude of the coefficient of output exceeded one in the case of the main crops of wheat and paddy in Punjab, indicating a more than proportionate rise in marketed surplus with the rise in output of these crops. In Bihar on the other hand, except for wheat, the coefficient of output was less than unity for all the major and minor crops. For the main staple crop of paddy, the coefficient was less than 0.3, indicating the subsistence nature of this crop. In the case of other variables, except for some commercial crops like cotton, potato and sugarcane, household size and marketed surplus had a significant negative relationship in the case of all major crops in both the states. The dummy variable for both market channels were significant in Bihar indicating higher market surplus for those households who sold their product in the regulated wholesale markets or in the local village market. In the case of prices, the value of marketed surplus and the value of output, the two values that also represented the economic status of the farmers, had significant and positive coefficients for most of the crops regressed in both the states, indicating a rise in the net price received by the farmers, with a rise in their economic status. In Bihar, the loan variable was another significant variable but with a negative sign, indicating the prevalence of credit and produce market interlinkages affecting the farmers' price negatively. However, with the widespread network of regulated markets in Punjab, such interlinkages were missing completely in that state and so the coefficient for the credit was insignificant for almost all crops in Punjab. Finally, the dummy variable for the market channel was insignificant in Punjab indicating no difference in price with respect to market channels, as more than 90 per cent of the output was marketed through regulated markets. In Bihar, the dummy variable for the first channel was insignificant indicating no price difference for those 10 per cent farmers who sold their product in the regulated *mandis*. That was probably the reason that farmers preferred to sell to the intermediaries.

Farm Investment in Assets Holding

As expected, there was a positive relationship between per household assets and farm size in both the states. However, the positive association did not hold in the case of per acre asset holdings. Small and marginal farmer invested a much higher amount per acre compared to large farmers in both the states. The value of assets per acre increased up to medium size farmers and declined for the large farmers in Punjab. In Bihar, marginal farmers' per acre investment was much higher compared to all other farm sizes. On an average, Punjab farmers invested a much higher amount on farming compared to Bihar and that was also reflected in their land productivity. In terms of per acre investment, Punjab farmers invested around double the amount of that of Bihar farmers. Comparing pre- and post-reform periods, our data supported the much discussed (Chand and Kumar, 2004) all-India phenomenon of rapidly increasing private investment in agriculture despite falling public investment during the nineties. From the pre- to post-reforms periods, investment for the selected households in Punjab increased by 78 per cent. The increase was even higher in the case of Bihar where investment increased by 125 per cent during the reform periods. Further, the bigger proportion of increase in investment came through small and medium farmers in Punjab and through marginal farmers in Bihar. Accounting for assets in which investment was made, tractors were the largest component of investment in both the states. A huge amount per household was spent on submersible tube wells in Punjab. The trends of investment were almost similar for all categories in both the states except for the case of marginal farmers in Punjab. In their case, it was apparent that these farmers in the wake of rising cost of operations were gradually shedding farming and thereby hardly making any new investment on the farm.

Major Problems Faced by the Respondents

In response to our survey on major problems faced, the respondents identified health, poverty, basic facilities, agriculture, employment and litigation as the major problems faced by them. Health problems and the lack of availability of suitable health centres or hospitals and doctors at reasonable rates was the biggest problem faced by the households in both the states. Around 40 per cent of the households in Punjab and 55 per cent in Bihar cited health problems as their biggest malaise, and the small and marginal farmers were not the only sufferers, but were also the large farmers. Shortage of basic facilities, stagnating land productivity and poverty or lack of money, especially during the slack season due to lack of supplementary earnings were the other major problems among the selected farmers in the two states. Employment opportunities for the educated youth and other supplementary employment opportunities were lacking in the peripheral areas according to the respondents in both the states. In the case of lack of infrastructure at the community or village level, a majority of the households in Bihar complained about the lack of physical infrastructure like roads, transportation, communication, electricity etc., while in Punjab it was the problem of social infrastructure like schools, community centres, health centres etc., which had become a stumbling block for the selected households. Thirty-six per cent of the respondents in Punjab and 41 per cent in Bihar were of the view that their plight could change by having sufficient acreage of land to till. However, a significant number of respondents in Punjab and Bihar wanted to opt out of farming and go for service, business or trading activities if the opportunity arose, to get a solution to their key problems. Noticeably, few respondents, only 10 per cent, viewed the government as a messiah who would deliver them from their suffering.

Changes in Income and Input Availability

On the question whether respondents' income had gone up or down during the last 10-year period since the globalisation process started in the agriculture sector, around 41 per cent of the households in Punjab and 58 per cent in Bihar were of the opinion that their income had gone up during this period. Around 25 per cent of the respondents in Punjab and 18 per cent in Bihar felt the opposite, while 34 per cent in Punjab and 24 per cent in Bihar were of the opinion that there was no significant difference in their income during the last 10 years. A majority of the households who felt strongly that their income had risen during this period were large and medium farmers in both the states. Among marginal farmers and landless labourers, only 20 to 30 per cent in Punjab and 40 to 50 per cent in Bihar opined that their income had increased. On the other hand, above 30 per cent of the marginal and landless farmers in Punjab and around 25 per cent in Bihar were disappointed by the policies of the last 10 years, which led to fall in their income during this period.

The respondents gave a high weightage to supplementary income, which they earned either through subsidiary activity along with the main farming activity or through permanent employment of some family members in these activities. Around 37 per cent of the households in Punjab and 32 per cent in Bihar viewed agriculture as the main activity, which made them better off than their other companions. However, 38 per cent of the respondents in Bihar were of the view that earnings from a regular job in the government or private sector provided them the much needed cushion that was not available through agricultural earnings. Similar views were expressed by Punjab farmers for the earnings they were getting from the side businesses like village shops or retail shops in the nearby town, *arhatia* or commission agent shop in the *mandi*, small mills in the village etc.

Almost all the respondents witnessed some marginal increase in the access to agricultural inputs during the last 10 years in both the states.

Around 90 per cent of the households belonging to all categories in Punjab accepted that they had access to institutional credit, chemical fertilisers, pesticides and HYV seeds. Moreover, access to these inputs somewhat increased during the reforms period. In Bihar, 73 per cent of the households were denied any access to institutional credit. Access to other inputs in Bihar was at par with that in Punjab. As in Punjab, respondents in Bihar also felt that their access to these inputs increased during the reform phase.

The major hurdles faced by the respondents in their agricultural operations were lack of quality inputs, improper price policy in the agriculture sector and lack of infrastructure. Around half of the selected households in Punjab and 70 per cent in Bihar had reservations about the available infrastructure in the agriculture sector, e.g., lack of proper banks and credit facilities, lack of markets or regulated *mandis*, *kachha* roads and improper means of transportation etc. In Punjab, a majority of the households were of the view that the prices of fertilisers, pesticides, tractors and implements have risen at a very rapid pace during the last decade whereas the price of various crops produced by them was increasing at a snail's pace. Their specific problem was the very high diesel prices that had turned agriculture uneconomic. Unfair trade in spurious pesticides and insecticides was another irksome problem faced by the respondents in both the states. Inadequacy of Government interventions in terms of procurement operations, research and extension activities and in the unforeseen events of droughts and floods was also highlighted by the respondents.

The most important activity among the social infrastructure villagers wanted to develop was a school, and a hospital/dispensary in the village. Laying of *pucca* or metalled road was also given a priority by a large number of respondents. Groupism, lack of cooperation among themselves and lack of proper guidance were given as the prime reasons for the lack of collective action on the part of the villagers in building the above infrastructure. Further, to a question on what help they desired from the government, 58 per cent of the respondents in Bihar sought an improvement in the physical infrastructure in their surrounding areas. In Punjab, a majority of the households wanted better access to agriculture inputs at cheaper prices. Improvement in financial and social infrastructure was also demanded by a large number of households in Punjab. In Bihar access to cheap and quality inputs was the second most important priority among the selected respondents.

Implications and Policy Thrust

The new buzzword of globalisation has multiple implications for the national economy as well as for the rural life style of millions. Globalisation has intensified interdependence and competition between economies in the world market. This is reflected in the interdependence in regard to trading in goods and services and in movement of capital. As a result, domestic economic developments are not determined entirely by domestic policies and market conditions. Rather, they are influenced by both domestic and international policies and economic conditions. It is thus clear that a globalising economy cannot afford to ignore the possible actions and reactions of policies and developments in the rest of the world while formulating and evaluating its domestic policy. This constrains the policy options available to the government, which implies loss of policy autonomy to some extent, in decision-making at the national level.

Understanding the current status of globalisation is necessary for setting the course for the future. However, each state differs in its economic set up and we cannot have a single set of policies to deal with all the states. As we have seen in the present case, agriculture in Punjab set the course of commercialisation way back in the early seventies with the beginning of the Green Revolution, and the state agriculture is now prospecting for a new phase of corporatisation of agriculture leading to the ascendancy of processing and value addition of farm products. On the other hand, agriculture in Bihar still lags behind with small farm orientation led by overwhelming subsistence concerns. The problems that the Punjab farmer is facing today relates to the question of sustainability, a consequence of intensive cultivation of land in the state. The Bihar farmer on the other hand is trapped in the web of backward infrastructure, low factor intensity resulting in low productivity and consequently low returns from the agriculture. Therefore, different policy actions are required to bring the farmers out of their dilemma especially the small and marginal ones who are at the edge of subsistence.

In the light of the findings of our study, the following suggestions and policy actions need to be put into action in the study area:

Bihar

Infrastructure

The biggest malaise Bihar is suffering from is the lack of infrastructure —not only physical infrastructure like roads, power, communication, and banking but also social infrastructure like health, education etc. The majority of the households expressed grave concern over the lack of infrastructure, which is a major hurdle in the development of the state. The table below shows the low status of Bihar in access to physical infrastructure.

Table 7.1

Status of Infrastructure in Bihar vis-à-vis Other States

	Road Density (per 100 sq km Area)	*Percentages of Villages Electrified*	*Average Area Served per Wholesaler Market (sq km)*
Bihar	77.86	72.18	392
Punjab	127.78	100.00	156
Tamil Nadu	158.78	100.00	434
Uttar Pradesh	86.77	79.40	456
West Bengal	85.00	78.10	318
India	74.93	86.50	451

Source: Thorat and Sirohi (2002a).

The creation of physical and social infrastructure comes in the ambit of public investment, as many of the basic amenities are public goods. Large irrigation schemes, roads and other communication systems, electrification, market development, flood control and drainage, etc., are some examples of public investment. Declining public investments in agriculture since the late eighties in the country and probably much more in the state has its possible adverse effects on the performance of agriculture. Although private investment in agriculture in the state has grown as was seen in Chapter 5, it cannot compensate for the decline in public investment and often involves inefficiencies. An example of this is the use of diesel pump sets, (which is high on cost and low on productivity) by the surveyed households due to lack of electrification in the state.

Highest priority should be given to increasing physical infrastructure, e.g., roads, communication, rural electrification etc. It has been shown empirically that the promotion of rural roads has the most beneficial impact on rural development. Investment in roads reduces rural poverty not only through growth in productivity but also through increased non-agricultural employment opportunities and higher wages, and providing the highest return in terms of both acceleration of growth and reduction in poverty. (Planning Commission, 2001; Fan *et al.*, 1999).

Irrigation

Irrigation is another area of major concern. In Bihar only 60 per cent of the cropped area is irrigated and that stands as the primary reason for low productivity of agriculture in the state. To raise opportunities for irrigation especially for the small farmers, the major focus should be on development of minor irrigation projects, which requires small investment as compared to major irrigation projects. As groundwater is abundant and cheap to extract, the emphasis should be on creating and maintaining capacity for lift and energising tube wells (Chandramohan *et al.*, 2001). Electric tubewells should be preferred to diesel tubewells since the former are efficient and less costly than the latter as was also seen in the case of Punjab. Thus, the procedure of lifting groundwater should be made more scientific and economical and its sustainability should receive the highest priority. Drainage is an equally important part of water management and there is a need for developing drainage projects in vulnerable areas. Irrigation, drainage and flood control need to be considered together in the context of each watershed (Kumar and Jha, 2003).

Consolidation of Holdings and Improving Tenancy Laws

It has been pointed out in the text that 80 per cent of the holdings in Bihar are below one hectare. The small holding size poses problems in the optimal use of land, application of machines and other investment in land improvement measures. Though the legislation for 'Consolidation of Holdings' was enacted in the state as early as in 1956, its implementation still has a major unfinished task. The main reasons for the slow pace of consolidation are inadequate financial outlays, shortage of manpower and, most important, the lack of political will. High priority needs to be accorded to the land consolidation programme. Further, along with consolidation of holdings, tenancy laws need to be revisited to secure the tenancy rights. The existing tenancy system supports sharecropping and tenancy in kind as was shown by our survey findings. However in the modern monetised economy, leasing in on fixed rent is bound to replace the sharecropping system. Further, the tenancy system continues to be concealed and informal with no security of tenure. Till 1986, tenancy rights were governed by the outdated Bengal Tenancy Act of 1885, which was amended in 1986. However, like other acts, the implementation of this act is also not effective.

Reforming Agriculture Marketing

Our findings on marketed surplus showed that in Bihar only 10 per cent of the output was sold through regulated *mandis* and a significant proportion of output was sold in the village itself through various intermediaries. Lack of metalled roads and proper regulated marketing network was the main reason for the farmers to opt out of sale in the village, even if it entailed lower prices, as was clear from the comparative price analysis of Punjab and Bihar. In the case of wheat and paddy, farmers in Bihar obtained prices lower than the MSP announced by the government. Marketing regulations in the state need drastic modifications. There are a number of areas where there are no regulatory provisions and no regulated *mandis* exist whatsoever. Therefore, if the plight of the farmers has to change and they are to be oriented to the market, a proper marketing network of regulated *mandis* in the proximity of each village or small town, along with legal and official framework in line with other advanced states like Punjab and Haryana is essential. It is also necessary to revisit the existing market act.

Another reason for farmers in Bihar getting lower prices is the unfavourable procurement policy of the Government of India in the case of major food grains. It was seen in Chapter 2 that 90 per cent of the wheat in Punjab, 80 per cent in Haryana and 39 per cent in Rajasthan was procured by the official agencies. Similarly in the case of rice, 84 per cent in Punjab, 59 per cent in Haryana and 66 per cent in Andhra Pradesh was procured by

FCI during TE 1999-2000. In contrast, procurement of wheat did not take place in Bihar and only 0.5 per cent rice was procured by the official agencies. This is happening despite substantial amount of grain is being offered for sale to these agencies. This step-brotherly behaviour of government intervention agencies further strains the situation as deserved support to the poverty-ridden small farmers in Bihar is denied to them. In the wake of increasing criticism and charges of Central procurement policy being biased and tilted towards states in which the farmers' lobby is much stronger, there is need to bring about reforms in the system (Acharya and Chaudhuri, 2001). Probably the role of procurement needs to be transferred to state agencies, which can do the job more equitably as pointed out in the Agricultural Policy 2000 (GoI, 2003).

Diversification towards High Value Crops and Enterprises

Marginal farmers dominate agriculture in Bihar with holding size less than one hectare, making large scale mechanisation impossible. The possibility of getting high returns from the traditional food grain production is meagre. In the given circumstances, a few lessons can be learnt from the recent experiments of modern high value cash crops grown by small farmers under the umbrella of contract farming. A few examples are cultivation of gherkin and baby corn in Karnataka, grapes in Maharashtra, tomato, potato and green chilly in Punjab and tomato in Haryana and raising broiler chicken in Tamil Nadu, (Singh and Asokan, 2003; Dileep *et al.*, 2002; Rangi and Sidhu, 2000 and 2003; Singh, 2005). It is observed in these cases that farmers, especially the small ones, can have better potential if they grow vegetable crops rather than continuing with traditional food crops. Under contract farming, small farmers get an assured buyer and are also able to get risk coverage through contracting agri firms. However, such a system can be more successful if these small farmers can come under the umbrella of a marketing organisation, which takes care of their production and post-harvest related activities in an integrated manner, and their other social and economic needs. This is fundamental for the success of any activity related to small farm agriculture. In addition to crop diversification, small farmers can diversify to high value enterprises like livestock, fisheries, bee keeping and so on, though again, for the success of these activities an effective support system is essential. Supply of inputs like seeds, planting material, credit, processing and marketing will need state support either directly or through incentives to the private sector.

Building up Institutions and R&D

Finally, agricultural research and development is an area of prime importance. There is underinvestment in agricultural research and development in the country during the last decade because of overspending on food, power and other agriculture related subsidies. The situation in Bihar is quite serious. The lone agricultural university in the state is finding it difficult even to pay regular salaries of its already pruned staff. The agricultural extension system is in complete disarray. Further, all institutions concerned with agriculture and rural development in the state, whether government or autonomous, have become practically non-functional. Political interference, bureaucratisation, nepotism and lack of professionalism can be attributed as reasons for the complete failure of the system (Kumar and Jha, 2003). Lack of coordination and integration of programmes of various departments results in dissipation of effort and inefficiency. The need of the hour is to take hard decisions to address these problems based on a sectoral rather than departmental perspective.

Punjab

Agriculture in Punjab is set on a different plane. Three decades of intensive agriculture now finds itself in a quandary. The sustainability of the present cropping pattern has become ecologically and economically costly. Productivity growth of important crops, especially paddy and wheat—the two most successful crops of Green Revolution, appears to have stagnated. On the other hand, the cost of production has started rising because of overuse of chemicals and machines and under use of family labour as was also observed by our survey data. It was pointed out in the beginning that in Punjab almost the whole of the cultivable area is already under the plough and therefore, possibilities of horizontal expansion are little. Given the fact that farmers are already having multiple crops and the cropping intensity is already touching the magic figure of 2, the vertical expansion of the cultivated area would also reach its limits very soon, except in the case of vegetable crops which in some cases are two-month crops and therefore farmers have the possibility of growing even four to five such crops in a year. Our survey results showed that although gross productivity in Punjab was more than twice that of Bihar, because of higher cost of production, the difference in net returns were much lower and output-input ratio in the case of large farmers in Punjab was less than Bihar for the same category of farmers. In other words, large farmers in Punjab were earning lower returns per unit of their agricultural inputs as compared to their counterpart large farmers in Bihar.

The extensive use of chemical fertilisers, imbalance in the use of NPK and overuse of pesticides as well as overexploitation of groundwater resources has resulted in degradation of soil and water resources in the state. The groundwater table is going down and soils are getting depleted in macro as well as micro-nutrients. In such a situation, efficient use of agricultural resources is indispensable to sustain productivity, production and income of the farmers in the state (Sidhu and Johl, 2002).

To break the deadlock, the pattern of production and distribution needs serious rethinking. The state cannot continue for long with the present cropping system of rice-wheat rotation. To divert farmers from the paddy-wheat rotation, there is a serious need to restructure the policies pertaining to the present way of input subsidisation and assured marketing through state procurement. The state government took a major step in achieving diversification for replacing rice and wheat in the *kharif* and *rabi* seasons by providing multiple options in terms of various coarse grains, pulses, oilseed crops, durum wheat and *basmati* rice through state-led contract farming with the help of the Punjab Agro Foodgrains Corporation (PAFC). Replacement of around 2.5 million acres of wheat and rice area by the above mentioned crops is being targeted by the year 2007. However, these efforts of the state government have remained by and large unsuccessful so far, and the ground reality does not seem to be changed (Kumar, 2005).

Apparently for a fairly long time, farmers in the state are used to an assured market for their produce and therefore their mindset is not prepared for facing market risks. Farmers of the state have to be trained to produce market based products if they have to raise their profitability from farming. The state government will have to play a big role in developing a niche market for the farmers' produce, developing commodity-specific marketing infrastructure and educating farmers about the demand based new products in which farmers can thrive-on in the near future.

Corporate-led contract farming is playing a significant role in evolving such a system and can help farmers diversify into new crops, especially in high value cash crops like fruits, vegetables, mushrooms, fisheries etc. It is a system under which agri-business firms/processors procure their raw material from the farmers at a pre-agreed price and the processors supply the growers with seed, other inputs, extension services and technical know-how. The system has already been introduced in the state for over a decade now. Several multinationals and indigeneous firms like Pepsi, FritoLay, Hindustan Lever, Chambal Agritech, A.M. Todd, Escorts, Tata Chemicals, Nijjer, DCM-Shriram, Mahindra ShubhLabh etc., are already procuring their material from the farmers under contract farming. The state government needs to promote this system on a larger scale by adopting certain policy actions, such as framing rules to provide a proper legal framework for the smooth functioning of the entrant firms as well as farmers in the contracting system. Certain tax concessions need to be given to attract a large number of firms especially in the area of fruits and vegetable processing. The government can impose certain conditions along with the provision of incentives to agri-business firms so that these firms are bound to involve the small and marginal farmers in their activities.

Further, it is desirable on the part of the state to integrate production with processing. Integration will help in controlling losses and increasing farmers' incomes, *vis-à-vis* reducing the cost of processing firms and upgrading the quality of their processed products. The degree of integration depends upon many factors such as type of crop, nature of commodity (perishability and seasonality), market conditions, technology availability etc. This approach should be encouraged to achieve quick diversification.

In addition to diversification towards high value cash crops, attempts should be made to diversify enterprises as well. In view of increasing job prospects for women, a new class of working families is fast emerging. Further, with increasing per capita income in the state as well as in the country, demand for packaged and ready-to-eat products is rising fast. Farmers of Punjab being entrepreneurial in nature would definitely take the advantage of this situation, if provided with such opportunities. A network of small scale rural enterprises for processing of agricultural produce could provide employment and income to the farmers and quality food material to the local and working population at relatively cheap prices, and primarily processed raw material to large urban industries. The opportunities can even be explored for export of such commodities as we can take advantage of the benefits of cheap labour and raw material as compared to many developed countries.

Last but not least, with the existing genotypes and production technology, there is still a gap of 2.0 to 2.5 tonnes per hectare of productivity of wheat and rice taken together, between the realisable potential and achieved levels (Sidhu and Johl, 2002). This has to be explored further through precision farming and improvement of the fertility of the soil. The state farmers being very forward-looking should be geared up to achieve precision by adoption of the recommended technological practices. The huge extension machinery in the state, research and development and agricultural universities should look beyond wheat and paddy and aim for higher productivity in the pulse crops which have a huge gap in the demand and supply in the state as well as in the country. Further, new and emerging crops should be the main focus of research for the longer period sustainability of the state agriculture.

References

Acharya, S.S. (1994). "Marketing Environment for Farm Products—Emerging Issues and Challenges", *Indian Journal of Agricultural Marketing* 8(2): 162-75.

Acharya, S.S. and D.P. Chaudhri (eds.), (2001). *Indian Agricultural Policy at the Crossroads: Priorities and Agenda*. New Delhi: Rawat Publications.

Bardhan, Pranab K. (1980). "Interlocking Factor Markets and Agrarian Development: A Review of Issues", *Oxford Economic Papers New Series* 32(1): 82-98. March.

Bathrick, David D. (1998). "Fostering Global Well-Being: A New Paradigm to Revitalise Agricultural and Rural Development", *IFPRI Discussion Paper 26*.

Bhalla, G.S. (1991). "Some Issues in Agricultural Marketing in India; Presidential Address delivered at the 5th National Conference of ISAM"; *Indian Journal of Agricultural Marketing* 5(3): 128-137.

———. (2005). "The State of the Indian Farmer", G. Parthasarathy Memorial Lecture delivered at the 88th Annual Conference of the Indian Economic Association, December 28. 2005, School of Economics, Andhra University, Vishakhapatnam, *Indian Economic Journal* 53(2).

Bhardwaj, Krishna (1974). *Production Conditions in Indian Agriculture*. Cambridge University Press.

Bisliah, S. (1978). "Decomposition Analysis of Employment Change under New Production Technology in Punjab Agriculture", *Indian Journal of Agricultural Economics* 33(2): 70-80.

Chadha, G.K. (1994). *Employment, Earnings and Poverty: A Study of Rural India and Indonesia*. Sage Publications.

———. (1999)."Non-Farm Employment in Rural Areas: How Well Can Female Workers Compete?", in T.S. Papola and Alakh N. Sharma (eds.), *Gender and Employment in India*. New Delhi: Vikas Publishing House.

Chand, Ramesh (2004). "India's National Agricultural Policy: A Critique", *Discussion Paper Series* No. 85/2004. Delhi: Institute of Economic Growth, Delhi University Enclave.

Chand, Ramesh and Parmod Kumar, (2004). "Determinants of Capital Formation and Agricultural Growth, Some New Explorations", *Economic Political Weekly* 39(52): 5611-6.

Chandramohan, C., Chitranjan and Sharat Kumar (2001). "Dynamics of Foodgrains Production", in Sharat Kumar and Praveen Jha (eds.), *Development of Bihar and Jharkhand: Problems and Prospects*. New Delhi: Shipra Publication. pp. 92-118

Dileep, B.K., R.K. Grover and K.N. Rai (2002). "Contract Farming in Tomato: An Economic Analysis", *Indian Journal of Agricultural Economics* 57(2): 199-210.

Fan, S., Peter Hazell and S. Thorat (1999). "Linkages Between Government Spending, Growth and Poverty in Rural India", *IFPRI Research Report 110*, Washington.

Ghuman, R.S. (2002). "WTO and Indian Agriculture: Crisis and Challenges—A Case Study of Punjab", in S.S. Johl and S.K. Ray (eds.), *Future of Agriculture in Punjab*. Chandigarh: Centre for Research in Rural and Industrial Development.

Gill, Sucha Singh (2002). "Agricultural Crop Technology and Employment Generation in Punjab", in S.S Johl and S.K. Ray (eds.), *Future of Agriculture in Punjab*. Chandigarh: Centre for Research in Rural and Industrial Development.

Government of India, (Various issues). *Commission for Agricultural Costs and Prices*. New Delhi: Department of Agriculture & Cooperation, Ministry of Agriculture.

———. (2003). *National Agriculture Policy 2000*. Ministry of Agriculture, Department of Agriculture and Cooperation.

———. (Various issues). *Agricultural Statistics At a Glance*. New Delhi: Directorate of Economics & Statistics, Department of Agriculture & Cooperation, Ministry of Agriculture.

———. (Various issues). *Area Production of Principal Crops in India*. New Delhi: Directorate of Economics & Statistics, Department of Agriculture & Cooperation, Ministry of Agriculture.

———. (Various issues). *Bulletin on Food Statistics*. New Delhi: Department of Agriculture & Cooperation, Ministry of Agriculture.

———. (Various issues). *Economic Survey*. New Delhi: Ministry of Finance.

Government of Punjab (1986). *Report of the Expert Committee on Diversification of the Punjab Economy*. Chandigarh.

Grewal, S.S. and A.S. Kahlon (1974). "Factors Influencing Labour Employment on Punjab Farms", *Agricultural Situation in India* 29(1): 3-5.

Harris, B. (1991). "Markets, Society and the State: Problems of Marketing under Conditions of Smallholder Agriculture in West Bengal", Report to WIDER, Helsinki, Oxford, Queen Elizabeth House.

Johl, S.S. (2002). *Johl Committee Report on Diversification of Agriculture in Punjab*. Chandigarh, Punjab.

Johl, S.S. and S.K Ray (eds.) (2002). *Future of Agriculture in Punjab*. Chandigarh: Centre for Research in Rural and Industrial Development.

Khusro, A.M. (1964). "Returns to Scale in Indian Agriculture", *Indian Journal of Agricultural Economics* 19(3-4): 51-80.

Krishna, Raj (1974). "Measurement of the Direct and Indirect Employment Effect of Agricultural Growth with Technological Change", in E.O. Edwards (ed.), *Employment in Developing Nations*. New York: Columbia University Press.

Kumar, Anjani and Dayanatha Jha (2003). "Agricultural Development in Bihar: Performance, Constraints and Priorities", (*Mimeo*). New Delhi: National Centre for Agricultural Economics and Policy Research.

Kumar, Parmod (1991). *New Technology and Income Distribution in Agriculture: A Case Study of Two Villages in Haryana*, (M.Phil. Dissertation), Jawaharlal Nehru University, New Delhi.

———. (1996). *Farm Size and Marketing Efficiency in Agriculture: An Analysis of Selected Markets and Crops in Haryana*, (PhD. Thesis), Jawaharlal Nehru University, New Delhi.

———. (2005). "Commercialization of Indian Agriculture and its Implications for Small and Large farmers", Paper Presented at FAO Conference on *Commercialization of Agriculture and Small Farmers,* Rome, 4-5 May, 2005.

———. (2006). "Contract Farming through Agri-Business Firms and State Corporation: A Case Study in Punjab", *Economic and Political Weekly* XLI(52): 5367-75. December 30.

Moore, J.R., S.S. Johl and A.M. Khusro (1973). *Indian Foodgrains Marketing*. New Delhi.: Prentice Hall of India Pvt. Ltd.

Nayyar, Deepak and Abhijit Sen (1994). "International Trade and the Agricultural Sector in India", in G.S. Bhalla (ed.), *Economic Liberalisation and Indian Agriculture*. New Delhi: ISID.

NCAER (ed.) (2001). *Economic and Policy Reforms in India*. New Delhi: National Council of Applied Economic Research.

Oberai, A.S. and Iftikhar Ahmed (1981). "Labour Use in Dynamic Agriculture—Evidence from Punjab", *Economic and Political Weekly* 16(13): A1-A4.

Olsen, W.K. (1993). "Distress Sales and Rural Credit: Evidence from an Indian Village Case Study", *IDS Bulletin* 24(3): 83-89.

Pandiyan Anand (1996). "Land Alienation in Tirunelveli District", *Economic and Political Weekly* 31(51): 3291-94.

Pingali, Prabhu L. and Sanjaya Rajaram (1998). "Technological Opportunities for Sustaining Wheat Productivity Growth Towards 2020", *IFPRI Vision 2020 Brief No. 51*. Washington D.C.: International Food Policy Research Institute, July.

Planning Commission (2001). *Approach Paper to the Tenth Five Year Plan (2002-2007)*. New Delhi: Planning Commission, Government of India.

Parihar, R.S. and D.S. Sidhu (1986). "Factors Affecting Labour Employment on Punjab Farms: An Economic Analysis", *Economic Affairs* 31(1): 9-19.

Pursell, G. (1999). "Some Aspects of the Liberalization of South Asian Agricultural Policies: How can the WTO Help?", in B. Blarel, G. Pursell and A. Valdés (eds.), *Implications of the Uruguay Round for South Asia: The Case of Agriculture*. Proceedings of a World Bank/FAO Workshop. New Delhi: Allied Publishers.

Raj, Krishna (1974). "Measurement of the Direct and Indirect Employment Effects of Agricultural Growth with Technical Change", in E.O. Edwards (ed.), *Employment in Developing Nations.* New York: Columbia University Press.

Raju, V. T. (1976). "The Impact of New Farm Technology on Human Labour Employment", *Indian Journal of Industrial Relations* 11(4): 493-510.

Rangi, P.S. and M.S. Sidhu (2000). "A Study on Contract Farming of Tomato in Punjab", *Agricultural Marketing* 42(4): 15-23.

Rangi, P.S. and M.S. Sidhu (2003). "Contract Farming in Punjab", *Productivity* 44(3): 484-91.

Rao, A.P. (1967). "Size of Holding and Productivity", *Economic and Political Weekly* 2: 1989-91.

Rao, C.H.H. (1965). "*Agricultural Production Functions, Costs and Returns*", New Delhi: Asia Publishing House.

———. (2003). "Reform Agenda for Agriculture", *Economic and Political Weekly* 38(7): 615-20.

Reddy, Narsimha (1997). "Economic Reforms, Entry of Corporate Sector in Agriculture and the Small Farmer Economy", in Y.K. Krishna Rao (ed.), *Liberalisation and the Small Farmer*. New Delhi: AIKS-ILO.

Roy, B.C. (1997). "Growth and Prospect of Fruit and Vegetable Processing Industry in India", *Bihar Journal of Agricultural Marketing* 5(3): 356-65.

Rudra, Ashok (1968). "More on Returns to Scale in Indian Agriculture", *Economic and Political Weekly* 3(26-28): 1041-44.

Selvaraj, K.N. and K.R. Sundaravaradarajan (1999). "Functioning of Informal Credit Markets and its Linkages: Evidence From Rural Credit Markets of South India", *Asia Pacific Journal of Rural Development* 9(1): 1-22.

Sen, A.K. (1964). "Size of Holdings and Productivity", *The Economic Weekly*, (Annual Number), February.

Sharma, Anil and Parmod Kumar (1999). "Market Intervention Schemes for Horticultural Commodities: Inefficiencies and Proposals for Reform", (*Mimeo*), NCAER.

Sharma, Anil, Rakhee Bhattacharjee and Parmod Kumar (2000). "Infrastructure Development Schemes for Agricultural Exports: An Evaluation", (*Mimeo*), NCAER.

Sidhu, R.S. and S.S. Grewal (1990). "Factors Affecting Demand for Human Labour in Punjab Agriculture: An Econometric Analysis", *Indian Journal of Agricultural Economics* 45(2): 1-10.

Sidhu, R.S. and S.S. Johl (2002). "Three Decades of Intensive Agriculture in Punjab, Socio-Economic and Environmental Consequences", in S.S. Johl and S.K. Ray (eds.), *Future of Agriculture in Punjab*. Chandigarh.

Singh, Gurdev and S.R. Asokan (2003). *"Contract Farming in India: Text and Cases"*, Ahmedabad: Centre for Management in Agriculture, Indian Institute of Management.

Singh, Sukhpal (2004). "Crisis and Diversification in Punjab Agriculture: Role of State and Agribusiness", *Economic and Political Weekly* 39(52): 5583-90.

———. (2005). *Political Economy of Contract Farming in India*. New Delhi: Allied Publishers Pvt. Ltd.

Srivastava, Ravi (2000). "Changes in Contractual Relations in Land and Labour in India", Tables 1 and 3, *Indian Journal of Agricultural Economics* 55(3): 253-82. July-September.

Thamarajakshi, R. (1999). "Agriculture and Economic Reforms", *Economic and Political Weekly* 34: 2293-95. August 14-20.

Thorat, S.K. and S. Sirohi (2002). Rural Infrastructure State of Indian Farmers, A Millennium Study. New Delhi: Ministry of Agriculture, Government of India.

———. (2002a). "Development of Rural Infrastructure in India: Trends and Emerging Issues", *State of Indian Farmer: A Millennium Study*. Ministry of Agriculture, Government of India. New Delhi: Academic Foundation.

Vaidyanathan, A. (1996). "Depletion of Groundwater, Some Issues", *Indian Journal of Agricultural Economics* 51(1-2). January-June.

———. (ed.) (2001). *Tanks of South India*. New Delhi: Centre for Science and Environment.

———. (2006). *India's Water Resources: Contemporary Issues on Irrigation*. New Delhi: Oxford University Press.

Vyas, V.S. (1998). "Changing Contours of Indian Agriculture", (*Mimeo*).

Appendix

Appendix I

Household Questionnaire

Project on
Economic Reforms and Small Farms

Sponsored by
Indian Council of Agricultural Research (ICAR)
Government of India

1.	District	[ ]	2.	Block	[ ]
3.	Village	[ ]	4.	HH. No	[ ]
5.	Religion	[ ]	6.	Caste	[ ]
5.	Respondent's Name....................		6.	Relation with the Head of Family	[ ]

Investigator's Name : ....

Signature and Date : ....

Supervisor's Name : ....

Signature and Date : ....

Institute for Human Development
IAMR Building
IP Estate
New Delhi – 110 002

1.DEMOGRAPHY

1a. Household Information

S. no. (ID No.)	Name	Relation with Head*	Sex (M/F)	Age (in Years)	Marital Status**	Educational Status #	Name of Occupation##			Industry Code $	Employment Status $$	Migrant Status $$$
							Main	Subsidiary				
								First	Second			
1	2	3	4	5	6	7	8	9	10	11	12	13

Codes:

* Relation with head of the household: Self-1; husband/wife-2; son/daughter-3; daughter/son-in-law-4; grand son/daughter-5; father/mother-6; brother/sister-7; mother/father-in-law-8; sister/brother-in-law-9; nephew/niece-10; uncle/aunty-11; other relatives-12; servant/employee/others-13.

** Unmarried-1; married-2; widow/widower-3; divorced/separated-4; others-5.

Illiterate-1; below primary or informal education-2; primary-3; middle-4; management/commercial school course-5; matric/high school/secondary-6; higher secondary/pre-university/intermediate-7; pre-graduation diploma or certificate below degree-8; degree equivalent to graduation level-9; postgraduate degree/diploma/certificate-10; technical degree (medical, engineering etc.)-11.

Write the exact name of occupation. Later provide occupation codes with the help of NCO code list given in the instruction sheet.

$ Only for main occupation. Industry code at two-digit level (See NIC code given with the instruction sheet).

$$ Self-employed-1; unpaid family worker-2; regular salaried-3; temporary salaried-4; ad hoc salaried-5; casual-6.

$$$ Resident-1; temporary (short-term) migrant (approx. 3 to 8 months out)-2; long-term migrant (out for 9 or more months)-3. Also note down number of months away from home during the last one year.

1b. Demographic Changes in the Household since 1991.

ID No.	Change of Occupation*			If Migrated, Place of Destination		Reasons for Change of Occupation*	Reasons for Change of Place**	Nature of Migration+	Amount of Remittances in 2001-02	Use of Remittances @
	Year	From	To	Place/ Distt/State	Rural/ Urban$					
1	2	3	4	5	6	7	8	9	10	11

* Self-cultivation-1; casual labour in agriculture-2; casual labour in non-agriculture-3; long-term attached labour-4; salaried-5; individual services (*yajmani*/caste occupation)-6; self-business based on agriculture allied activities-7 (animal husbandry; poultry, aquaculture, etc.)-7; small business/trade/construction-8; big business/trade/construction-9; collection/collection of fodder/cutting wood from jungle and selling-10; others (specify)-11.

$ Rural-1; urban-2.

** More earnings-1; better scope of work-2; sent by friend/relative/contractor-3; quarrel/tension in village-4; got job-5; others (specify)-6.

+ Temporary-1; permanent-2; circular-3

@ Payment towards debt-1; treatment of family member-2; purchase of land-3; construction/renovation of house-4; marriage expense-5; purchase of consumer durable-6; day-to-day expenses-7; lend on interest-8; purchase of machinery/commercial vehicle-9; others (specify)-10.

2. OPERATIONAL HOLDING, LAND TRANSACTION, LAND AND OTHER ASSETS

2a. Land

Nature of tenure	Distribution of area by source of irrigation (Acres)					Total	Rate of Rent, in Case Leased-in/out
	Only Canal	Canal+ Tube Well (Electric/Diesel)	Only Electric Tube Well	Only Diesel Tube Well	Rainfed Area		
1	2	3	4	5	6	7	8
1. Total owned land:							
a. Under cultivation							
b. Cultivable waste							
c. Non-cultivable							
2. Leased-in land							
3. Leased-out land							
4. Mortgaged-in land							
5. Mortgaged-out land							
6. Land encroached by you							

2b. Details of Land Leased-in

	Land 1	Land 2	Land 3	Land 4
1	2	3	4	5
1. Type of lease*				
2. Area (acres)				
3. Rent paid during last year (total value of cash and kind)				
4. Terms of payment of cash rent**				
5. Input costs shared by landlord(%)(a)				
(b)				
(c)				
6. Any other				
7. Share in output				
8. Loan received from landlord (Rs.)				
9. Since when cultivating this plot?				
10. Since when cultivating for the same landlord?				

* Share cropping-1; fixed rent in cash-2; fixed rent in kind-3; both-4; against labour-5; others-6.

** All cash paid in advance for the whole year-1; advance payment in two instalment of six months-2; all the amount paid after the harvest-3; any other term-4.

2c. Land Transactions since 1985

	Year of Transaction	Actual Rate Rs/Acre	Socioeconomic Character of Buyer			If Land Sold, Purpose	If Bought, Source of Finance
			Occupation	Class*	Caste		
1	2	3	4	5	6	7	8
1. Amount of land sold							
a.							
b.							
c.							
2. Amount of land bought							
a.							
b.							
c.							

* If peasant/farmer use standard category of large (>10 acres), medium (5-10 acres), small (2.5-5 acre) and marginal (0-2.5 acre) category; for non-farmer use estimates of assets.

2d. Ownership of Productive Assets

Sl. No.	Assets	Number	Year of Purchase	Bought Old/New	Current Value Rs	Asset Disposed off		Disposal Value	Reason for Disposal*
						1991-95	After '95		
1	2	3	4	5	6	7	8	9	10
1.	Tractor								
2.	Trolley								
3.	Barrow								
4.	Tiller								
5.	Plank								
6.	Threshing machine								
7.	Combine harvester								
8.	Other reaper (specify)								
9.	Pump set diesel								
10.	Pump set electric- Submersible Non-submersible								

contd...

...contd...

Sl. No.	Assets	Number	Year of Purchase	Bought Old/New	Current Value Rs	Asset Disposed off		Disposal Value	Reason for Disposal*
						1991-95	After '95		
1	2	3	4	5	6	7	8	9	10
11	Bullock cart								
12	Fodder Chaffer— Manual Power driven								
13	Sewing machine								
14	Spray pump								
15	Storage bin								
16	Poultry sheds								
17	Dairy sheds								
18	Honey bee box								
19	Any other (i)								
20	(ii)								

* To clear old debt-1; marriage-2; other social ceremonies-3; medical expenses-4; litigation-5; to purchase better asset/implement-6; exchanged with old one-7; others (specify)-8.

2e. Livestock

Types of Animals	Number Owned in 1991	Number Owned currently	Present Value of Stock	Reason for the Difference*
1	2	3	4	5
1. Cow (*desi*)				
2. Cow (hybrid/cross-bred)				
3. Female-buffalo				
4. Female-goat				
5. Young cattle				
6. Bullocks/draught animals				
7. Poultry				
8. Other animals (a)				
(b)				
9. Value of leased-out animals				

* Purchased-1; sold-2; died-3; gifted-4.

2f. Other Household Assets

S.No	Assets	Number		Old/ New when Purchased	Current Value Rs	Asset Disposed off		Disposal Value	Reason for Disposal*
		1995	2002			1991-95	After '95		
1	2	3	4	5	6	7	8	9	10
1.	House								
2.	*Palang*								
3.	Cot								
4.	Table								
5.	Chair								
6.	Almirah								
7.	TV (colour/BW)								
8.	Radio								
9.	Deck								
10.	VCR/VCP								
11.	Cycle								
12.	Motorcycle/scooter								
13.	Car/jeep								

contd...

...contd...

S.No	Assets	Number		Old/New when Purchased	Current Value Rs	Asset Disposed off		Disposal Value	Reason for Disposal*
		1995	2002			1991-95	After '95		
1	2	3	4	5	6	7	8	9	10
14.	Pressure cooker								
15.	Electric fan								
16.	AC/Air cooler								
17.	LPG								
18.	Electric stove								
19.	Kerosene stove								
20.	Fire hearth (*Chullah*)								
21.	Microwave oven								
22.	Any other (specify)								
	a.								
	b.								
	c.								
	d.								

* To clear old debt-1; marriage-2; other social ceremonies-3; medical expenses-4; litigation-5; to purchase better asset/implement-6; exchanged with old-7; others (specify)-8.

3. Housing Status

Item	1991	1995	Current Status
a. Details of homestead land*			
b. Total area of homestead land (sq mts)			
c. Total covered area of the house (sq mts)			
d. Whether house electrified [Y/N]			
e. If not, then source of light in the house**			
f. Source of drinking water***			
g. Distance of the source of drinking water			
h. Place of defecation #			
i. Main source of fuel ## (if more than one source give % for the different sources)			
j. Distance travelled for collection of fuel			

* Owned but inherited-1; Owned but bought-2; provided by government-3; *gair majarua* (government land without allotment)-4; landlord's land-5; others (specify)-6.

** Lantern-1; *dhigri*-2; kerosene petromax-3; gas petromax-4; others (specify)-5.

*** Own hand pump/tube well-1; motorised hand pump-2; public hand pump/tube well-3; tap in dwelling-4; own protected dug well-5; unprotected dug well-6; public unprotected dug well-7; public protected dug well-8; public tap-9; pond, river, stream-10; shared hand pump-11; others-12.

Septic tank-1; pit latrine-2; covered dry latrine-3; community latrine-4; in the field-5; others-6.

Wood-1; coal-2; kerosene oil-3; hay/leaves-4; cow dung cake-5; agricultural waste (stalks)-6; *gobar* gas plant-7; liquid petroleum gas-8; electricity-9; others-10.

4. INDEBTEDNESS

4a. Details of debt

Details of Debt	Debt 1	Debt 2	Debt 3	Debt 4
1	2	3	4	5
1. Cash or kind*				
2. Source**				
3. Purpose of loan***				
4. Year of loan taken				
5. Amount of loan taken (in Rs.)				
6. Condition of loan#				
7. Rate of interest (per annum)				
8. Total debt outstanding (Rs.)				

* Cash-1; kind-2.

** Institutions (other than government programme)-1; traders-2; money lenders-3; (*Ahartia)* Commission agent-4, landlord/employer-5; friends/relatives-6; own caste men-7; upper caste men-8; lower caste men-9; government programme-10; others (specify)-11.

*** Daily consumption-1; marriage-2; health treatement-3; purchase of land-4; purchase of machinery-5; purchase of means of transport-6; purchase of domestic consumer durables-7; crop loan-8; construction of house-9; litigation-10; purchase of livestock-11; police diary-12; working capital in business/non-agricultural enterprises-13; any other (specify)-12.

On interest-1, labour service-2, mortgage of land-3, mortgage of other properties-4; others (specify)-5.

4b Do you do wage work/work as tenants for those to whom you are indebted? (Y/N) [ ]

4c Is there any cooperative credit society in your village? (Y/N) [ ]

4d If yes, is any person of your family a member of such a society? (Y/N) [ ]

4e Is there any informal credit society/self-help group in your village? (Y/N) [ ]

4f If yes, is any person of your family a member of such a group? (Y/N) [ ]

4g Do you have any account in a bank/post office/any other similar institution? (Y/N) [ ]

4h Do you have any stocks, bonds, shares or any other similar things? (Y/N) [ ]

4I Do you have Life Insurance Policy? (Y/N) [ ]

5. CULTIVATION AND LABOUR USE
(To be asked to those who cultivate own or tenanted land)

5a Cropping Pattern and Yield during Last Six Months
(Area in acres, output in quintals and value in Rs.)

Sl. No.	Crops	Total Area		Irrigated Area	HYV area	Physical Output		Value of Main Product	Value of Byproduct
		1995	Now			Main	Byproduct		
1	2	3	4	5	6	7	8	9	10

5b. Details of Advanced Seed Used

Sl. No.	Name of Crop	Seed Variety*	Name of Seed	Source of Information**	Source of Technology Provider**	Whether Approved by Government
1	2	3	4	5	6	7

* Hybrid-1; Biotech (Genetically Modified)-2; HYV-3; Any other (specify)-4.

** Agricultural Universities-1; Private Companies-2; Seed Vendors-3; Other Farmers-4; Any other (specify)-5.

5c. Input Costs in Cultivation

Input Costs		Name of Crop								Total Cost
1		2	3	4	5	6	7	8	9	10
1. Ploughing charges, if hired										
2. Seed (if purchased)										
3. Fertiliser										
4. Irrigation (if bought, canal/ tube well)										
5. Pesticides/insecticides										
6. Harvesting	a. Hired machine									
	b. Hired labour									
7. Threshing (if hired)										
8. Total fuel cost of tractor, thresher, pump set etc.										
9. Maintenance cost of tractor, thresher, pump set etc.										
10. Hired labour cost										
11. Any other cost (specify) a.										
b.										
12. Total										

5d. Sale/Marketing of the Crop

Sale/marketing of crops	Crops							
	Crop-1	Crop-2	Crop-3	Crop-4	Crop-5	Crop-6	Crop-7	Crop-8
1	2	3	4	5	6	7	8	9
1. Sold where* a.								
b.								
2. Who purchased** a.								
b.								
3. Quantity sold (quintals)								
4. Price/qtl (or unit specify)								
5. Total income (Rs.)								
6. Transportation cost								
7. Marketing fee								
8. Any other cost								
9. If stored, amount (in qtls)								
10. Period (in months)								

* Wholesale market-1; local market-2; villagers (directly)-3; cooperative-4; government (levy)-5; government agencies-6; sold to merchant against debt/advance-7; free selling at farm gate-8; pre-arranged contract-9; others (specify)-10.

** Landlord-1; private merchants-2; government agencies-3; villagers-4; others (specify)-5.

5e. Cropping Pattern and Labour Use in Last Six Months

Crop Name	Agricultural Operation	Family Labour			Attached Labour		Hired Labour			
		Male	Female	Child	Local	Migrant	Local-M	Local-F	Local-C	Migrant
1	2	3	4	5	6	7	8	9	10	11
1.	a. Sowing									
	b. Caring									
	c. Harvesting/threshing									
	d. Marketing									
2.	a. Sowing									
	b. Caring									
	c. Harvesting/threshing									
	d. Marketing									
3.	a. Sowing									
	b. Caring									
	c. Harvesting/threshing									
	d. Marketing									
4.	a. Sowing									
	b. Caring									
	c. Harvesting/threshing									
	d. Marketing									

contd...

...contd...

Crop Name	Agricultural Operation	Family Labour			Attached Labour		Hired Labour			
		Male	Female	Child	Local	Migrant	Local-M	Local-F	Local-C	Migrant
1	2	3	4	5	6	7	8	9	10	11
5.	a. Sowing									
	b. Caring									
	c. Harvesting/threshing									
	d. Marketing									
6.	a. Sowing									
	b. Caring									
	c. Harvesting/threshing									
	d. Marketing									
7.	a. Sowing									
	b. Caring									
	c. Harvesting/threshing									
	d. Marketing									
8.	a. Sowing									
	b. Caring									
	c. Harvesting/threshing									
	d. Marketing									

6. LIVESTOCK PRODUCTION AND EMPLOYMENT

6a. Input Costs for Livestock during Last One Year

Inputs	Cost in Rs	Proportion Supplied from	
		Home	Market
1	2	3	4
1. Fodder			
a) green			
b) Dry			
c) Concentrates			
d) Others			
2. Medicines			
3. Others			
a)			
b)			
c)			

6b. Production and Income from Livestock during Last One Year

Items Produced	Value of Production during Last One Year (Rs)	Value of Quantity Sold (Rs)	Marketing of Product*	Transportation Cost (Rs)
1	2	3	4	5
1. Milk				
2. *Ghee*				
3. *Paneer*				
4. Cow dung				
5. Egg				
6. Poultry (meat)				
7. Goat				
8. Pig				
9. Hide skin				
10. Others.(specify)				
a.				
b.				
c.				

* Within village-1; at collection centre-2; in the town-3.

6c. Labour Use in Livestock Monthly Average during Last One Year (Man-days)

Activities	Family Labour				Permanent Labour			Hired Labour		
1	2	3	4	5	6	7	8	9	10	11
	M	F	MC	FC	M	F	C	M	F	C
1. Collection of fodder and its cutting										
2. Normal looking after of animals										
3. Collection of cow dung & making dung cakes										
4a. Others (specify)										
b.										
c.										
d.										

7. Other Self-Employment (except Agriculture): Employment and Earnings in Last One Year

	Occupa-tion-1	Occupa-tion-2	Occupa-tion-3	Occupa-tion-4
1	2	3	4	5
a. Family Labour (man-days)				
1. ID No.				
2. ID No.				
3. ID No.				
4. ID No.				
b. Utilisation of labour				
1. Total Hired labour (man-days)				
2. Expenditure on labour (Rs)				
c. Other input cost (Rs)				
d. Gross income (Rs)				
e. Net income (Rs)				

8. REGULAR NON-FARM EMPLOYMENT

8a. Employment and Wages during Last One Year

	ID No.	ID No.	ID No.
1	2	3	4
1. Work status*			
2. Occupation			
3. Industry			
4. Working since how long? (month/year)			
5. Period of contract**			
6. Days worked			
7. Any other work?			
8. Average monthly income (Rs.)(including cash and kind)			
9. Other income (value of gifts etc. received)			
10. Total			
11. Mode of payment***			
12. Works in village or outside$			
13. If outside, where#			
14. Travelling costs in commuting (Rs.)			

* Government/semi-government job-1; large private company-2; small business/industry-3; family business/industry-4; others-5.

** 1-monthly cash payment; 2-monthly kind payment; 3-annual cash payment; 4-annual kind payment; 5-daily cash wage; 6-daily kind wage; 7-others.

***: Contract for indefinite period-1; contract for definite period-2; permanent-3.

$ Within village-1; outside-2.

\# Neighbouring village-1; far off village-2; nearby town-3; others (specify)-4.

9. CASUAL LABOUR IN AGRICULTURE AND NON-AGRICULTURE

9a. Casual Labour (in Last Six Months)

Name	ID No.	Industry	Occupation	No. of Days	Total Wages (Rs.)	Mode of Payments*
1	2	3	4	5	6	7
June						
July						
August						
September						
October						
November						

* Only cash-1; only kind-2; cash and kind both-3.

10. OTHER INCOME AND EXPENDITURE

10a. Other Sources of Income and Heads of Expenditure during Last One Year

Source/Item	Amount (Rs.)
1. Income	
a) Rent from leasing out of land	
b) Rent from leasing out of agricultural implements	
c) Rent from house	
d) Interest on lending	
e) Old age/widow pension	
f) Other pensions	
g) Remittances received	
h) Any other receipts	
2. Expenditure (non-consumption)	
a) Repair of house	
b) Repair/servicing/maintenance of agricultural implements	
c) Repair/servicing/maintenance of other productive assets	
d) Other heads of expenditure	
(i) *Malgujari*/tax	
(ii)	

10b. Expenditure Pattern of Household

Items	Average Monthly Expenditure $ (Rs.)	Items	Average Monthly Expenditure $ (Rs.)
1. Food grains*		11. Savings / lending	
2. Non-food grains**		12. Marriage/ *Shradh* etc.	
3. Intoxicants		13. Others (specify)	
4. Education		a)	
5. Fuel & electricity		b)	
6. Medical expenses		c)	
7. Transport & communication		d)	
8. Clothing		e)	
9. Recreation***		f)	
10. Loan repayment			

$ For question no. 1 to 5 collect the expenditure data for the last 30 days only. For rest of the item collect the expenditure data for last 12 months and commute it for average of one month.

* Expenditure on cereals and pulses only.

** Expenditure on vegetables, milk, meat, fish, eggs, fruits (dry & fresh) etc.

*** Expenditure on cultural activities, festivals etc.

10c. Do you get two full meals (sufficient food) throughout the year? (Y/N) [ ]

10d. If not, in which months you do not get sufficient food:

June	July	August	September	October	November

10d. During months in which you do not get sufficient food, how do you cope up with shortage? [ ]

(Take loan-1; catch fish/rat/crab etc.-2; near starvation-3; sometimes starvation-4; take meal only once-5; begging-6; others (specify)-7).

10e. Has there been any change in your own economic condition since 1995? [ ]

(Code: Improved-1; deteriorated-2; remained the same-3).

10f. Has there been any change in village life since 1995? [ ]

(Code: Improved-1; deteriorated-2; remained the same-3).

11. ABSOLUTE AND RELATIVE DEPRIVATION

11a. Do you feel that your family has not sufficient food for the whole of year? If yes, give reasons.

11b. Have you faced any deprivation other than food insufficiency? If yes, explain.

11c. What are the main difficulties you and your family faced during the last year?

11d. What is the most important thing your household lacks?

11e. What is the suggestion for amelioration of your condition?

11f. What is the most important thing your village lacks?

11g. Perceptions and changes

(i) Since 1995, has your income gone up (1),down (2), or stayed about the same (3) ? [ ]

(ii) What is the most important thing which makes some people better off than others in this village?

(iii) What can villagers do collectively for the overall development of the village?

(iv) If collective action is possible, then reasons for villagers not acting so?

12. AGRICULTURE AND RELATED POLICY (For cultivators only)

12a. Do you get the Following Things Adequately for Your Cultivation?

Items	Adequately*		Reason for Inadequacy**	
	1995	At Present	1995	At Present
1. Credit from bank, cooperative, government				
2. Chemical fertiliser				
3. Pesticides				
4. HYV seeds				
5. Hybrid seeds				
6. Genetically modified (biotech) seeds				
7. Bio-fertilisers/pesticides				
5. Any other (specify)				

* Yes-1, No-2; somewhat adequate-3.

** Inadequate at the moment-1; out of reach-2; prices are high-3. not available in the fair shop-4; not in nearby place-5; do not get from the government-6; others (specify)-7

12b. (i) Have you ever got any help or advice in your agriculture work from government extension workers (like VLW)? (Y/N) ☐

(ii) If yes, how many times during last one year?

(iii) What kind of help did you receive?

12c. In your view what are the main difficulties, which you are facing in raising your agricultural production?

12d. What kind of help do you want from the government?

Appendix II

Village Schedule

Economic Reforms and Small Farms: Implications for Production, Marketing and Employment

1. VILLAGE PROFILE

Village	Block	District	State	Nearest Urban Centre

2. INFORMATION TO BE COLLECTED FROM BLOCK/CIRCLE OFFICE

2a. Basic data on Population

	2001	1991
1	2	3
1. Number of households		
2. Number of males		
3. Number of females		
4. Total population		
5. Number of literate males		
6. Number of literate females		
7. Number of scheduled castes		
8. Number of scheduled tribes		
9. Number of backward castes		
10. Number of upper castes		

2b. Occupational Structure (Census Classification)

	2001		1991	
	Male	Female	Male	Female
1	2	3	4	5
1. Cultivators				
2. Agricultural labourers				
3. Livestock, forestry etc.				
4. Mining, quarrying etc.				
5. Household industry				
6. Other manufacturing				
7. Construction				
8. Transport				
9. Services				
10. Other workers				
11. Total main workers				
12. Total marginal workers				
13. Total non-workers				

2c. Land Use Pattern (Acre/Hectare)

	2001	1995	1991
1	2	3	4
1. Total reported area			
2. Forest			
3. Not available for cultivation			
4. Permanent pasture & other grazing land			
5. Land under misc. tree, crops & groves			
6. Cultivable waste land			
7. Fallow other than current fallow			
8. Current fallow			
9. Net sown area			
10. Area sown more than once			
11. Total cropped area			

2d. Net Area Irrigated by Different Sources (Area in Acres)

	2001	1995	1991
1	2	3	4
1. Canal			
2. Tube well (electric)			
3. Tube well (diesel)			
4. Well			
5. Tank			
6. Unirrigated			
7. Other			

2e. Livestock (Number)

	In-milk	Dry	Young Stock Female	Young Stock Male	Adult Male	Total
1	2	3	4	5	6	7
1. Indigeneous cow						
2. Cross-bred cow						
3. Buffalo						
4. Indigeneous sheep						
5. Cross-bred sheep						
6. Goat						
7. Pig						
8. Local poultry						
9. Improved poultry						
10. Other animals						
a.						
b.						
c.						

2f. Number of Households across Different Land Class

	2001		1991	
	Number of HH	Total Area (in Acres)	Number of HH	Total Area (in Acres)
1	2	3	4	5
1. Landless				
2. 0-1 acre				
3. <2.5 acre				
4. 2.5-5.0 acre				
5. 5-10 acre				
6. 10-20 acre				
7. > 20 acre				
8. Total				

2g. Cropping Pattern from Village Records

Season	Crop Name	Total Area (in Acre)		Irrigated Area (in Acre)	
		2001	1995	2001	1995
1	2	3	4	5	6
	1. Cereals				
	a. Paddy				
	b. Wheat				
	c. Maize				
	d Other cereals				
	i.				
	ii.				
	iii.				
	iv.				
	2. Pulses				
	a. *Arhar*				
	b. Gram				
	c.				
	d.				
	3. Oilseeds				
	a. Mustard				
	b. Linseed				
	c.				
	d.				
	3. Cash crops				
	a. Sugarcane				
	b. Jute				
	c. Cotton				
	d.				
	e.				
	4. Fodder crops				
	5. Fruits & vegetables				
	a.				
	b.				
	c.				
	d.				
	6. Any other (specify)				
	a.				
	b.				

Information to be collected from Knowledgeable Persons in the Village

3. DISTRIBUTION OF HOUSEHOLDS ACCORDING TO SIZE CLASS OF LAND HOLDING

3a. Number of Households from Different Castes Groups

Caste	Landless	Marginal (<2.5 acre)	Small (2.5-5.0)	Medium (5-10)	Large (> 10)	Total
1	2	3	4	5	6	7
1.						
2.						
3.						
4.						
5.						
6.						
7.						
8.						
9.						
10.						

3b. Number of Households involved in Different Occupations

	Landless	Marginal (<2.5)	Small (2.5-5.0)	Medium (5-10)	Large (> 10)	Total
1	2	3	4	5	6	7
1. Growing of vegetables						
2. Growing of flowers						
3. Commercial forestry						
4. Growing fruits						
5. Milk vendors						
6. Livestock activity						
7. Fish farming						
8. Poultry keeping						
9. Bee-keeping etc.						
10. Shop-keeping/trading						
11. Rice/wheat/oil/ sugarcane crushing						
12. Other agro-processing activities						

4. INFRASTRUCTURE FACILITY

4a. Electricity

1. Is the village electrified? (Y/N) :
2. If Yes, since when (Year)? :
3. Electricity availability (average number of hours per day) :
4. If electricity was available
5. 5 Years ago, for how many hours per day :
6. 10 years ago, for how many hours per day :

4b. Electricity Connection by Land Class

Number of connection	Land Class Group				
	Marginal	Small	Medium	Large	Total
	1	2	3	4	5
a. Domestic					
b. Agricultural					
c. Commercial/business					
d. Industrial					
e. Others (specify)					

4c. Transport, Communication, Credit and Market

	Within Village (Yes=1, No=0)	If no, Name the Nearest Village/Town	Distance from Village (km)
1	2	3	4
1. Road connectivity			
2. Railway connectivity			
3. Telephone			
4. Post office			
5. Cooperative credit society			
6. Regional rural bank			
7. Commercial bank			
8. Agricultural market			
9. Vegetable market			
10. Fruit Cooperative marketing society			
11. Vegetable Cooperative society			
12. Dairy Cooperative society			
13. Fisheries Cooperative society			
14. Input shop			
a. Seed: wheat			
b. Rice			
c. Vegetables			
d. Fruits			
15. Chemical fertilisers			
16. Biofertilisers			
17. Biopesticides			
18. Veterinary centre			
19. Repairs of agricultural machinery			
19. Any other (specify):			
a.			
b.			
c.			
d.			

4d. Education Facilities

Type	Place	Distance	Type of Approach Road*
1	2	3	4
1. Primary school (boys/co-ed.)			
2. Primary school (girls)			
3. Middle school (boys)			
4. Middle school (girls)			
5. High/higher secondary school (boys)			
5. High/higher secondary school (girls)			
6. Intermediate college			
7. Degree college			
6. Religious school			
7. Non-formal centre			
8. Other educational facilities (specify)			
a. I.T.I.			
b.			
c.			
d.			

* Type of approach road: *Pucca* - (1); *Kuchha*-(2); Semi-*Pucca*-(3).

4e. Health Facilities

Type	Place	Distance (in km.)	Type of Approach Road	Period Accessible	
				By Foot	By Jeep
1	2	3	4	5	6
1. Primary health sub-centre					
2. Primary health centre					
3. Hospital/dispensary					
4. Private qualified allopathic doctor					
5. Maternity/child care centre					
6. Family planning clinic					
7. Chemist/medical shop					
8. Others (specify)					
i.					
ii.					
iii.					

4f. Other Facilities

Type	Place	Distance (in km.)	Type of Approach Road	Period Accessible	
				By Foot	By Jeep
1	2	3	4	5	6
1. Block HQ					
2. Nearest town					
3. Nearest bus stop					
4. Nearest regular market					
5. Nearest rail station					
6. Police station					
7. *Gram panchayat* office					
8. Fair price shop					
9. Other general shop					
10. Petrol pump					
11. Others (specify):					
a.					
b.					
c.					

5. VILLAGE ORGANISATIONS

Are organizations of the following type active in the village?

Name of the Organisation	Type*	Functions**	Member	Location	Level of Activity***
1	2	3	4	5	6
A. Cooperative					
i. Credit					
ii. Agricultural inputs					
iii. Production of *khadi*					
iv. Marketing					
v. Dairy cooperative					
vi. Others (specify)					
i.					
B. Organizations/unions					
i. Workers organisations/unions					
ii. Farmers organisation					
iii. Voluntary organisations					
i.					
iv. Religious/caste					
v. Political organisations					
vi. Cultural organisations					
vii. Youth *mandal*					
viii. Women *mandal*					
ix. Self-help groups					
x. Other (Specify)					

* Government-1; non-government organisation supported-2; privately managed group-3; political-4; other (specify)-5.

** Awareness generation-1; village development work-2; political movement-3; profit making-4; helping farmers/labourers-5; cultural activities-6; solving disputes-7; other (specify)-8.

*** Normal-1; good-2; very good-3; not active-4; near to dead-5.

6. CROP OPERATION, LABOUR USE AND WAGES

6a. Crop Operations and Timings

(Specify name of months of start and end of different operations)

Crop Operations 1	Duration* 2	Crop 1 3	Crop 2 4	Crop 3 5	Crop 4 6	Crop 5 7	Crop 6 8	Crop 7 9	Crop 8 10
1. Name of crop									
2. Land preparation, bunding, ploughing	S								
	E								
3. Sowing	S								
	E								
4. Transplanting	S								
	E								
5. Weeding/hoeing A. Ist weeding/hoeing	S								
	E								
B. 2nd weeding/hoeing	S								
	E								

contd...

...contd...

Crop Operations 1	Duration* 2	Crop 1 3	Crop 2 4	Crop 3 5	Crop 4 6	Crop 5 7	Crop 6 8	Crop 7 9	Crop 8 10
6. First irrigation	S								
	E								
7. Second irrigation	S								
	E								
8. Harvesting	S								
	E								
9. Post harvest operations specify for each crop (husking, retting etc.)	S								
	E								
10. Average yield (excluding harvest cost)	S								
	E								
11. Harvest share	S								
	E								
12. Minimum and maximum price at which crops were sold last year	S								
	E								

* S and E stand for month of start and month of end of particular operation.

6b. Irrigation Charges (Per Acre)

Season	Canal	Staff Tube well	Private Electric	Private Diesel
1	2	3	4	5
1. *Kharif*				
2. *Rabi*				
3. *Zaid Kharif*				
4. *Zaid Rabi*				
Five Years Ago				
5. *Kharif*				
6. *Rabi*				
7. *Zaid Kharif*				
8. *Zaid Rabi*				

6c. Average Cost of Machines

Name of Crops	Tractor per Acre	Thresher/Harvester per Quintal	Sprayer per Hour	Combined Harvester
1	2	3	4	5
1. *Kharif*				
2. *Rabi*				
3. *Zaid Kharif*				
4. *Zaid Rabi*				
Five Years Ago				
5. *Kharif*				
6. *Rabi*				
7. *Zaid Kharif*				
8. *Zaid Rabi*				

6d. Labour Use

Name of Crops	Use of Human Labour	
	Average Man-days Used per Acre	Wage Paid per Day
1	2	3
a. *Kharif/Bhadai*		
(i)		
(ii)		
(iii)		
(iv)		
(v)		
b. *Rabi*		
(i)		
(ii)		
(iii)		
(iv)		
(v)		
(vi)		
(vii)		
c. *Zaid Kharif*		
(i)		
(ii)		
(iii)		
(iv)		
d. *Zaid Rabi*		
(i)		
(ii)		
(iii)		
e. Others		
(i)		
(ii)		
(iii)		

6e. Prevailing Wages

Task		Type of payment made in general? Daily wage...1 Piece rate (per unit land).....2 Piece rate (crop share)...3 Others (specify) .4	What are the prevailing wages for casual labour for the following tasks? For piece rate work, estimate conversion into daily wage rate in areas where multiple forms of payment are used, report wage rates in terms of the most prevalent form of payment Record the total number of meals/snacks received each day in column and the value of daily cash and payments in kind in the second and third column. Do not include the value of meals in the columns for payments in kind.								
			Male			Female			Child (10-14 Years)		
TYPE	Average Wage Rate		No. of Meals	Cash	In Kind	No. of Meals	Cash	In Kind	No. of Meals	Cash	In Kind
1	2	3	4	5	6	7	8	9	10	11	12
1. Prevailing agri. wages											
2. Ploughing											
3. Hoeing											
4. Weeding											
5. Paddy transplanting											
6. Harvesting of wheat											
7. Harvesting of paddy											
8. Harvesting of grams											

contd...

...contd...

1	2	3	4	5	6	7	8	9	10	11	12
9. Harvesting of pigeon pea											
10. Harvesting of maize											
11. Cane-cutting											
12. Digging of potatoes											
13. Hashing of wheat											
14. Winnowing of wheat/paddy											
15. Construction											
16. Govt. programmes											
17. Other Skilled work											
a.											
b.											
c.											
d.											

7. LAND AND OTHER PRODUCTIVE ASSETS

7a. Distribution of Cultivable Land by Quality Type

No.	Type of Land	Area (Acres)	Irrigated	Present Value (per Acre)
1	2	3	4	5
1.				
2.				
3.				
4.				
5.				

Source/Respondents:

7b. State the Number of Households that gone out of Cultivation across Size of Land-holding since 1995

	Landless	Marginal (<2.5)	Small (2.5-5.0)	Medium (5-10)	Large (> 10)	Total
1	2	3	4	5	6	7
1. Present main occupational status						
a.						
b.						
c.						
d.						
e.						
f.						
g.						
h.						

7c. Capital and Livestock Assets

	Number of Cultivators			
	Ownership		Utilisation	
	Owning	Since when*	Using	Since when*
1	2	3	4	5
1. Private pump set/boring				
2. Tractor				
3. Power tiller				
4. Harvester combine				
5. Thresher				
6. Seed drill				
7. Improved cattle livestock				
8. Pucca grain storage				
9. Improved implements				
(a)				
(b)				
(c)				
10a. Other (specify)				
b.				
c.				

* 1-Last few years; 2-during 1996-1999; 3-during 1991-1995; 4-before 1991.

8. MARKETING OF AGRICULTURAL PRODUCE

8a. The Climatic and Other Conditions for Agriculture Last Year

8a(i). How was the climatic condition last year?
(About normal-1; better than normal-2; worse than normal-3)

8a(ii) If Different from normal, give reasons and causes:

8b. Marketing of Crops

Crop (Specify Season)	Average Yield (qtls/acre)	Sales*	Methods of Sale**
1	2	3	4
1.			
2.			
3.			
4.			
5.			
6.			
7.			
8.			
9.			
10.			
11.			
12.			
13.			
14.			

*Sales: Categories as (1) almost all sold; (2) more sold than kept; (3) fair amount sold but more kept; (4) only a little sold; (5) more sold.

**Methods: Categories as (1) direct sale in wholesale market, mention place; (2) intermediary from outside village; (3) intermediary from inside village; (4) direct sale to other villagers; (5) sales through cooperative; (6) pre-arranged contract; (7) other (specify).

8c. Crop Prices for Main Marketed Crops

Crop (Specify Season)	Wholesale Prices (Mention Units)							
	May 03	Code	Nov 02	Code	May 96	Code	Nov 95	Code
1	2	3	4	5	6	7	8	9
1.								
2.								
3.								
4.								
5.								
6.								
7.								
8.								
9.								
10.								

Note: The information has to be collected from market/block office. In case they are not possible, they should be collected from concerned knowledgeable persons.

Code: Sold below MSP-1; at MSP-2; higher than MSP-3; No MSP-4; any other (specify)-5.

8d. Innovation

(i) New agricultural practices being used since 1995

Name of the Crop	Class Group[1]	Seed Variety[2]	Agricultural Practices[3]	Mechanical Technology[4]	Irrigation Practices[5]	Marketing Practices[6]
1	2	3	4	5	6	7
1.						
2.						
3.						
4.						
5.						
6.						
7.						
8.						

1 Class group (cultivator): marginal farmer-1; small-2; medium-3; large-4.

2 Seed variety: hybrid-1; genetically modified-2; tissue culture (horticulture)-3; any other (specify)-3.

3 Agricultural practices: organic farming-1; any other (specify)-2.

4 Mechanical technology: introduction of new machine (specify)-1; Improvement of existing machine-2; others (specify)-3.

5 Irrigation practices: drip irrigation-1; sprinkle irrigation-2; pipe irrigation-3; water harvesting technology-4; any other (specify)-5.

6 Marketing practices: online marketing-1; forward buying-2; contract selling to government-3; contract selling to cooperatives-4; contract selling to MNCs-5; Contract selling to other Private organisations-6.

9. TENANCY

9a. Questions for Households Leasing out Land

On land which is leased, which arrangements are found in the village: (put 1, 2 etc., in order of importance, if arrangement is not found)

1. No leasing out :
2. Crop-Share (annual) :
3. Crop-share (seasonal) :
4. Fixed annual rent (crop) :
5. Fixed annual rent (cash) :
6. Fixed seasonal rent :
7. Longterm fixed rent :
8. Other (specify) :

9b.Under Each of these Arrangements, What are the Most Common Practices

Type of Tenancy (Specify)	Type of Tenancy		
1	2	3	4
1. Rent/share of landlord			
2. Period same plot normally given			
7. Share in inputs by landowner (bullock/tractor)			
4. Seed or seed costs			
8. Irrigation costs			
6. Fertiliser costs			
7. Whether labour supplied by tenant to landowner?			

9c. 1. Over the past 10 years has tenancy increased, decreased or remained the same?

2. If increased or decreased, then by what proportion

a) Proportion of leasing in households :

b) Proportion of leased in area :

9d. Questions for Tenants

On land which is leased in, which arrangements are found in the village: (put 1, 2 etc., in order of importance, if arrangement is not found)

1. No leasing out :
2. Crop-share (annual) :
3. Crop-share (seasonal) :
4. Fixed annual rent (crop) :
5. Fixed annual rent (cash) :
6. Fixed seasonal rent :
7. Long-term fixed rent :
8. Other (specify) :

9e. Under Each of these Arrangements, What are the Most Common practices

Type of Tenancy (specify)	Type of Tenancy		
1	2	3	4
1. Rent/share of landlord			
2. Period same plot normally given			
3. Period same landowner			
4. Share in inputs by landowner (bullock/tractor)			
5. Seed or seed costs			
6. Irrigation costs			
7. Fertiliser costs			
8. Whether labour supplied by tenant to landowner?			

9f. 1. Over the past 10 years has tenancy increased, decreased or remained the same?

2. If increased or decreased, then by what proportion

a) Proportion of leasing in households :

b) Proportion of lesed in area

:

9g. Distribution of Leasing out of Land by Households across Size Class of Holding

	Marginal	Small	Medium	Large	Total
1	2	3	4	5	6
1. Number of households					
2. Area (in acre)					
3. No. leasing out on					
a. Fixed annual rent					
b. Crop share					
c. Any other (specify)					
4. Area leasing out on					
a. Fixed annual rent					
b. Crop Share					
c. Any Other (specify)					

10. COMMON PROPERTY RESOURCES

10a. What are the main Common Property Resources (CPRs) in the village? Their uses and encroachments on them ?

Main CPRs	Are they in use?	If Yes, Which Class of People?	Are the Poor People Prohibited in Using?	Any Encroachment:	By Whom?*
1	2	3	4	5	6
a. Forest					
b. Village pond					
c. Pasture					
d. *Ahar/pain*					
e. *Garmazrua* land					
f. Other (specify)					

* Landless-1; marginal farmer-2; small farmer-3; medium-4; large-5.

11. LABOUR MARKET INFORMATION

11a. Labour within Village (Only Casual Labour)

Sector of Employment	Local labour						In Migrant Labour					
	Persons Involved		Average Days of Employment Per Worker		Wage Rate (in Rs)		No. of Immigrant Labour		Average Days of Employment of Per Worker		Wage Rate (in Rs)	
	M	F	M	F	M	F	M	F	M	F	M	F
1	2	3	4	5	6	7	8	9	10	11	12	13
1. Agriculture												
2. Livestock												
3. Forestry												
4. Fishing												
5. Manufacturing and processing in Household Industry												
6. Manufacturing & Processing in non-Household Industry												
7. Construction												
8. Transport												
9. Others (specify)												
i.												
ii.												
iii.												

11b. Attached Labour in Agriculture

Details of Contract	Halwaha	Charwaha	Kamia	Others (Specify)
1	2	3	4	5
1. Total number in village				
2. Type of contract (No.) a. Seasonal b. Annual c. Long-term				
3. Types of work performed				
4. Average hours of duty per day				
5. Daily wage rates (Rs)				
6. Annual payments (Rs)*				
7. Condition of leaving employment**				
8. Whether land is provided***				
9. If land is provided, area in acres				
10. Land produce is sharable with landlord $				

* Include all kinds of gifts/clothing/footwear etc.

** No condition–1; free after one season or one year–2; can leave only after debt repayment–3.

*** yes–1; no–2.

$ Yes–1; no–2 (also mention proportion of share).

11c. Do some people who usually live in the village and go to work outside the village daily? If yes, give details.

Sl. No.	Type of Work	Where	Distance (in km)	Months	Numbers	Approximate Monthly Earnings
1	2	3	4	5	6	7
1.						
2.						
3.						
4.						
5.						
6.						
7.						
8.						

11d. Do some people who usually live in the village, go to work outside village for longer period? If yes, give details.

Sl. No.	Type of	Where	Distance	Months	Number	How Arranged*	Approximate Monthly Earnings
1	2	3	4	5	6	7	8
1.							
2.							
3.							
4.							
5.							
6.							
7.							
8.							

* Through an outside contractor–1; through local intermediary of contractor–2; through relatives/friends–3; villagers forming their own gang–4.

12. OTHER INFORMATION

12a. In the last year, how many households/persons have left the village? In the last five years how many have left the village excluding those counted above:

Tola	Last Year	Preceding Five Years	Total
1	2	3	4
1.			
2.			
3.			
4.			

12b. What are the household and non-household enterprises running in the village?

Type of Enterprises	Number of Enterprises		Main Products	What are the Input Types	Explain the Types of Support from Government/ *Panchayat*
	Household	Non-Household			
1	2	3	4	5	6
1.					
2.					
3.					
4.					
5.					
6.